Y0-DOK-464

Benchmarking for
Effective Network Management

McGraw-Hill Computer Communications Series

In order to receive additional information on these or any other McGraw-Hill titles in the United States please call 1-800-822-8158. In other countries, contact your local McGraw-Hill representative.

Benchmarking for Effective Network Management

Dr. Kornel Terplan

McGraw-Hill, Inc.
New York San Francisco Washington, D.C. Auckland Bogotá
Caracas Lisbon London Madrid Mexico City Milan
Montreal New Delhi San Juan Singapore
Sydney Tokyo Toronto

Product or brand names used in this book may be trade names or trademarks. Where we believe that there may be proprietary claims to such trade names or trademarks, the name has been used with an initial capital or it has been capitalized in the style used by the name claimant. Regardless of the capitalization used, all such names have been used in an editorial manner without any intent to convey endorsement of or other affiliation with the name claimant. Neither the author nor the publisher intends to express any judgment as to the validity or legal status of any such proprietary claims.

Library of Congress Cataloging-in-Publication Data

Terplan, Kornel.
    Benchmarking for effective network management / by Kornel Terplan.
        p.   cm.
    Includes bibliographical references and index.
    ISBN 0-07-063638-9
    1. Computer networks—Management.   2. Benchmarking (Management)
I. Title.
    TK5105.5.T47   1995
    004.6—dc20   95-6614
                                                                    CIP

Copyright © 1995 by McGraw-Hill, Inc. Printed in the United States of America. Except as permitted under the United States Copyright Act of 1976, no part of this publication may be reproduced or distributed in any form or by any means, or stored in a data base or retrieval system, without the prior written permission of the publisher.

hc 1 2 3 4 5 6 7 8 9 DOC/DOC 9 9 8 7 6 5

ISBN 0-07-063638-9

The sponsoring editor for this book was Jerry Papke. The executive editor was Joanne M. Slike. The managing editor was Lori Flaherty. Theresa Cunningham was the manuscript editor. The director of production was Katherine G. Brown. This book was set in ITC Century Light. It was composed in Blue Ridge Summit, Pa.

Printed and bound by R.R. Donnelley & Sons Company, Crawfordsville, Indiana.

Information contained in this work has been obtained by McGraw-Hill, Inc. from sources believed to be reliable. However, neither McGraw-Hill nor its authors guarantee the accuracy or completeness of any information published herein and neither McGraw-Hill nor its authors shall be responsible for any errors, omissions, or damages arising out of use of this information. This work is published with the understanding that McGraw-Hill and its authors are supplying information but are not attempting to render engineering or other professional services. If such services are required, the assistance of an appropriate professional should be sought.

MH95
0636389

To Terezia Terplan, whose support allowed me to devote the necessary time to this publication.

Contents

Acknowledgments

I would like to thank Lori Flaherty and Theresa Cunningham for the excellent copyediting work and Adam Szabo and Alan Bookmiller for all the illustrations. Without their help I would not have been able to submit my text to the publisher on time.

I would also like to thank Jill Huntington-Lee, Brandywine Consulting, and Russ Lisle, Lynx Technologies, for their review of the manuscript and their excellent comments.

I am particularly grateful to Jerry Papke, senior editor at McGraw-Hill, for his support of this book.

Introduction

Budget pressures force network managers to improve the efficiency of their communication networks. The key to improvement lies in processes, instruments, and the human resources of network management. In most cases, network managers are not aware of their relative position to their peer operations: are they stronger or weaker than industry average? In other cases, decisions about outsourcing network management functions are under consideration. Nobody, however, really knows the value of network management to the network or the value of the network to the corporation. Network management benchmarks can answer both questions.

Thanks to a growing practice known as *benchmarking*, it is becoming increasingly common for companies to get together and compare notes on which problems are prevalent and how to solve them. Theoretically, these companies leave with a stronger sense where their operations stand relative to others in and out of their industry. The mediator role is usually taken by consultant companies. Benchmarking helps determine the actual definition of accomplishments and how companies can match or exceed the best in the business. The consideration of "best practices" in other industry operations can now be applied to communications, which could also be important when considering insourcing instead of outsourcing.

The book assists benchmarking by identifying benchmarking phases, procedures, and instruments. The examples, based on practical cases, can accelerate the implementation of benchmarking techniques. This book targets management processes, instruments, standards, and human resources for benchmarking a network. This book does not, however, target actual network benchmarking. Systems management is now converging with network management, but very few examples are available about the effectiveness of the individual components of benchmarking. Benchmarking systems management is outside the scope of this edition.

The purpose of the book is to improve the performance of network management processes, the use of network management instruments, and the allocation of human resources and instruments by highlighting gaps and comparing different businesses to each other. It can also be used to implement the benchmarking process with the process and instrument allocation forms, questionnaires, performance indicators, job profiles, and list of expected skill levels. In this case, both clients and benchmarking companies are the target audience. The book can also be used by the client to control the quality of work of the benchmarking companies who are conducting the benchmarks.

Chapter 1 defines the targets of network management benchmarks that can be used as comparisons with the best practice and when outsourcing. The typical targets are status review, strengths, and weaknesses of the businesses under consideration. The benchmarking process has several risks because of the high impact of interview results, documentation inaccuracy, and false interpretation of indicator results. This chapter also includes examples for benchmarking results.

Chapter 2 introduces the phases of benchmarking. The *data gathering phase* consisting of studying documentation, interviewing staff and observing operations, is the key phase. *Data consolidation* is the interpretation of measured/gathered data. In most cases, unsophisticated tools can be used here. Gaps can be identified by comparing key indicators with industry averages, assuming the benchmarking supplier maintains benchmarking data for many clients. The final phases address the recommendations for filling gaps, supporting full or partial outsourcing, restructuring responsibilities, changing process flows, using other instruments, or changing the training curriculum.

Chapter 3 focuses on the critical success factors of network management, including processes, instruments, standards, and human resources. This chapter offers in-depth data collection forms for network management functions, instruments, standards, and job descriptions. Skill matrices are introduced for each network management function.

Chapter 4 is helpful for all clients who want to define indicators and quantify performance. This chapter covers questionnaires and generic indicators, all measurable and quantifiable. Particular interest is focused on the service quality of the client service point—a single point of contact for all clients served by the communication network. This chapter also covers network and network management costs; the cost of network ownership is answered by analyzing hardware, software, communications, infrastructure, and personnel expenses.

Chapter 5 presents data collection techniques for benchmarking indicators. Collection techniques, strategies, and tools are introduced along with their applicability. The results from questionnaires, interviews, observations, and logs are presented. Examples for various companies that represent small and large networks and network management organizations are shown in this chapter.

Chapter 6 addresses the most creative activities of interpreting the data collected in Phase 1 of the benchmarking project. The principal emphasis is on the quantitative correlation between network management functions and personnel, network management functions and instruments, and personnel and instruments. This segment also addresses whether and how network management functions can be assigned to human resources.

Chapter 7 quantifies the value of the network to the corporation, and the value of network management to the communication networks. Comparisons are offered between industry averages and the special indicators of the corporation conducting the benchmark. In the comparisons, both ideal and realistic targets are used.

Chapter 8 identifies the basics of outsourcing network management functions, instruments, and personnel. Outsourcing criteria, outsourcing partners, and outsourcing alternatives are discussed. This segment also includes the estimation of timeframes and expenses connected to outsourcing decisions.

Chapter 9 addresses the recommendations for the corporation after evaluating the results of the benchmarking project. These recommendations can include actions, timeframes, and costs for improving the overall effectivity and efficiency of operations. The recommendations are usually presented to top management after the written report has been submitted.

Chapter 10 predicts the future of network management benchmarking, including the most likely suppliers of such services.

Overview of Benchmarking

Budget pressures force network managers to improve the efficiency and effectiveness of the use of communication networks. The key to improvement lies in processes, instruments, standards, and the human resources of network management. In most cases, network managers are clearly aware of the value of their operation, but they want to know if they are stronger or weaker than what is average in the industry. Network management audits and benchmarks help answer this question.

1.1 Network Management Benchmarking

Network management means deploying and coordinating resources to plan, operate, administer, analyze, evaluate, design, and expand networks to meet service-level objectives at all times at a reasonable cost with reasonable capacity. Basically, network management benchmarks can be used for three purposes (see Figure 1.1):

- highlighting gaps in network management processes, instrumentation, standards, and human resources
- offering comparisons with industry averages or best practices
- preparing outsourcing decisions

In the first two cases, indicators are needed to quantify performance. This book uses data collection techniques, including forms and interviews, to evaluate which network management processes are supported, which in-

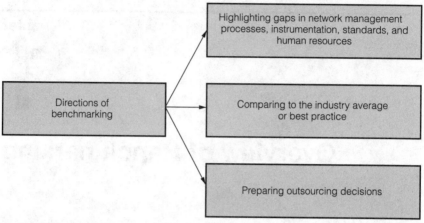

Figure 1.1 Directions of benchmarking

struments are used, which management protocols have been implemented, how human resources are assigned to processes and instruments, and what the skill levels of the management team are. To get feedback about the networking environment, investments, and process details, three different questionnaires are used. They can be filled in prior to or during benchmarking. To quantify and compare performance of network management, a number of benchmarking indicators are used grouped around generic, organizational, process-specific, and cost indicators. The results can be used for individual companies to identify areas of improvement or compare performance with the industry average or best practice. Usually, clients are interested in doing both.

The third target is preparing outsourcing decisions. A mutual interest exists for both the client and outsourcer that they quantify the value of the network, its management, and its human resources. Network management benchmarks can easily be incorporated into the "diligency" segment of the outsourcing process.

Thanks to the growing acceptance of audits and benchmarking, it is becoming increasingly common for companies to get together and compare notes on how they solve problems. Theoretically, these companies come away from benchmarking with a stronger sense of their operational position relative to others in and out of their industry. The mediator role is usually supported by consulting companies.

Auditing and benchmarking help determine which accomplishments really exist or can be achieved and also give companies the chance to match or exceed the best in the business. These activities are now an integral part of the Total Quality Management (TQM) program, based on ISO-9000 or TQM. ISO-9000 (Hutchins 1994) offers practical guidelines for how to improve the quality of conducting business and has become a condition of business from the European point of view. TQM has similar goals from the American point of view.

This chapter shows a number of different examples using various indicators. The goal of this book is to demonstrate detailed and quantified benchmarking results from previous studies. No logical sequence or interrelationship exists between the examples in this chapter. Most of the indicators are self-explanatory; in some cases, short explanations are given. In most cases, however, Chapter 4, which discusses quality indicators, is required to fully understand the interpretation of indicators. The samples in this chapter relate to the management of both data and voice networks.

1.2 A Benchmarking Example

Figure 1.2 shows an example of benchmarking results, using four indicators to compare the client's performance with the industry average. The indicated area of the graph in the figure clearly shows the company where and how to improve. The indicators are the following:

Indicator 1. The number of proactive fault detections compared to all faults (*proactive* means the use of intelligent, continuous monitoring to recognize deteriorating equipment or facilities as early as possible).

Industry average   60 percent
Client company   40 percent

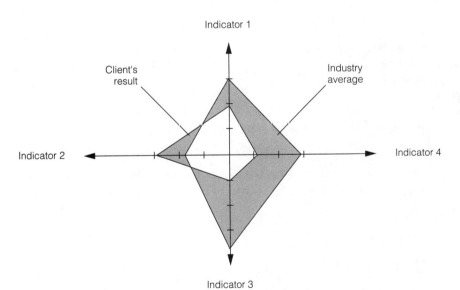

Figure 1.2 Typical benchmarking star

Indicator 2. The percentage of multiple responsibilities for network management functions (multiple responsibilities can hinder efficiency, however, because of the finger pointing that occurs when either serious or subtle faults are identified).

Industry average   35 percent
Client company   60 percent

Indicator 3. The percentage of referral in the fault-resolution process compared to all faults (usually, multiple referrals, which means multiple persons, results in a longer resolution process).

Industry average   80 percent
Client company   20 percent

Indicator 4. The percentage of correctly completed change request forms compared to all change requests. Note that incomplete forms delay the change execution process and can hinder successful execution).

Industry average   60 percent
Client company   20 percent

The basic idea of auditing and benchmarking is to identify several organizations that represent best practices in the same functional and service areas where the interested company needs improvement. After identifying comparative companies, several meetings are arranged, research conducted, questions answered, and potential improvements highlighted to the mutual benefit, in most cases, of the participants.

An additional goal is to make outsourcing decisions. Network management audits and benchmarks provide a good estimate on the value of existing network management processes, instruments, and personnel. Besides the actual value, top management receives detailed reports on the following items:

- inventory of network components
- inventory of network management instruments
- organizational structure of network management
- skill levels, ages, and salary ranges for network management personnel
- allocation of functions and instruments to network management personnel
- statement of mission or list of objectives

Benchmarking helps observe dynamic changes in the performance of certain indicators. Figure 1.3 shows the dynamic of one special indicator, the percentage of proactive fault detections compared to all faults. This indicator starts at lower than industry average at the initial benchmarking study and overtakes the industry average by the second benchmarking study.

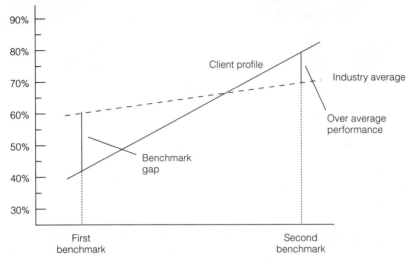

Figure 1.3   Dynamic changes between two benchmarking studies

This dynamic can be influenced by the level of investment in network management processes, instruments, and education. To demonstrate the power of benchmarking, several examples with various indicators are shown.

1.3   Workload Composition

To plan for growth and changes in the corporation, it is interesting to compare the workload composition with other corporations. The network-related workload usually includes:

- file transfer
- interactive transaction traffic
- electronic mail (E-mail)
- voice
- fax
- video

If necessary, further details can be measured and reported using application monitors and performance databases. Figure 1.4 shows the distribution for Company A and B. Both companies are in the same industry. Similar graphics can be generated to compare companies with the industry average. The industry average might show in this particular case more traffic for voice and

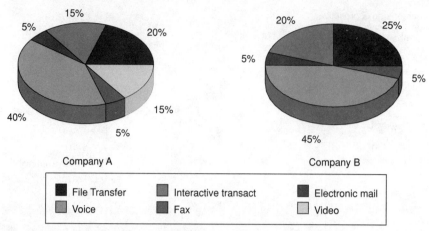

Figure 1.4 Comparison of network-related workload

less traffic for file transfers. These results were derived from approximately 40 surveys conducted by the author in various countries and industries.

1.4   Network Budgets

Communication budgets are growing in almost all corporations, but the growth rates are mostly unable to match expectations. Thus, corporations are very interested in comparing their results related to spending and the network management budget. Costs can be broken down accordingly:

- hardware
- software
- communications
- personnel
- infrastructure

Figure 1.5 shows the results for a sample year, displaying the distribution for Companies A and B and the industry average. These results were derived from approximately 40 surveys conducted by the author in various countries and industries.

Many corporations are uncertain of the methods used to select and invest in the various network management functions. Comparison with the industry average could be very helpful in reviewing their own strategy. Investments can be grouped around key network management functional areas. This example subdivides the expenditures into network operations, network administration, customer support, and planning and design.

Figure 1.6 displays the deviations for Companies A and B compared to the industry average. Company A is in deficit; it spends less than the industry average. The opposite is true for Company B, which spends more than the industry average. Again, these results were taken from surveys conducted by the author.

1.5 Size of Network Team

Because of the high costs of human resources, corporations are very interested in the size of the network management teams within other corporations, as well as the ratio among managers, supervisors, and subject-matter experts. The overall basis is often the number of managed objects. In this respect, it is very difficult to find sound industry averages. Figure 1.7 shows two examples. The examples are shown in detail in Table 1.1, which shows the internal resources. In other chapters, the full size of the team is evaluated, including external resources of consultants and contracted labor.

1.6 Success Factors of Change Management

Moves, additions, and changes cannot be eliminated or reduced in today's and future networks. The ultimate goal is to properly control and manage these processes. One of the quality indicators is how successfully these processes are handled by the network management organization.

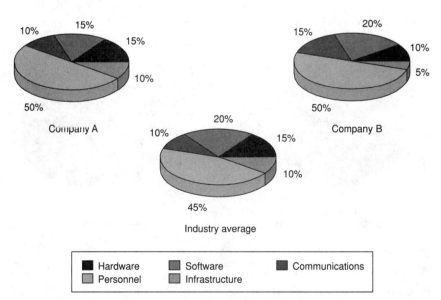

Figure 1.5 Distribution of spending network management budget

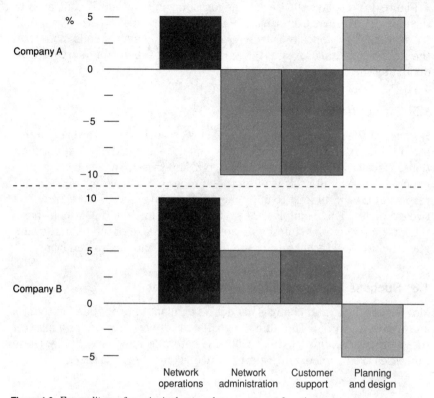

Figure 1.6 Expenditures for principal network management functions

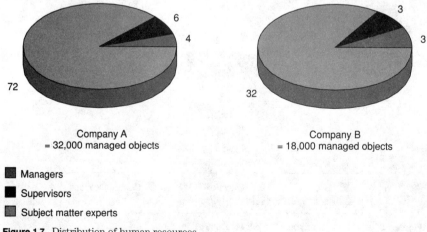

Figure 1.7 Distribution of human resources

TABLE 1.1   **Distribution of Human Resources**

		Company A	Company B
Managed objects	=	32,000	18,000
Managers	=	4	3
Supervisors	=	6	3
Subject matter experts	=	72	32

Shown in Figure 1.8 from the surveys conducted, the author compares two companies using the following subindicators based on the percentage of changes:

- completed successfully on time
- completed successfully but delayed
- rejected because of incomplete request forms
- failed during execution

1.7   Fault Resolution

The speed of fault resolution is crucial to resuming service to the users and is significantly impacted by the right mix of instruments and the right level of education for the operating staff. Companies are still in doubt about how to justify additional expenditures. Figure 1.9 shows the efficiency of a fault-resolution process using the number of referrals as indicators. Each additional referral means a substantial time delay in resolution.

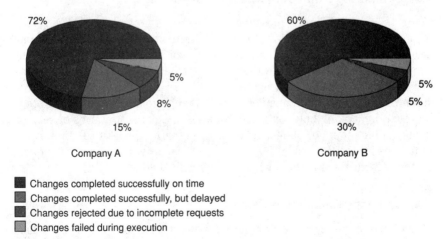

■ Changes completed successfully on time
□ Changes completed successfully, but delayed
■ Changes rejected due to incomplete requests
□ Changes failed during execution

Figure 1.8   Success factors of change management

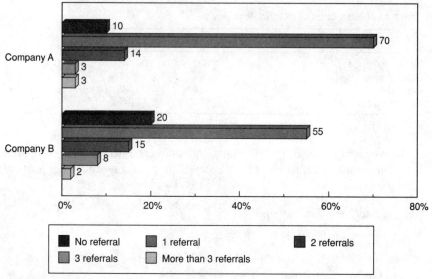

Figure 1.9 Efficiency evaluation of fault management

By examining the numbers, the two companies compared seem very different. Company B shows a lower percentage for a single referral than Company A. Adding up one referral and no referrals, however, balance the overall performance. In the more-than-one-referral category, both companies are almost equivalent. These indicators were developed from the evaluation of thousands of actual trouble tickets.

1.8 Comparing Number of Devices per Kilobit

Users, telecommunication providers, and vendors prefer very high-level indicators to sell or justify investments. The next few indicators are concerned with bandwidth and choices in voice networks. Investing in transmission and end-user devices can be evaluated by comparing the number of devices per kilobit of bandwidth. Figure 1.10 shows the overall results for a number of companies. These results might be misinterpreted, however, by looking only for a high number of devices. In terms of service quality, a high number of devices is not necessarily the best solution. Prior to final interpretation, the utilization level of facilities and the service quality for personnel using the devices must be checked.

1.9 Company Call-Minute Costs

Phone charges today are very competitive and telephone services are fast becoming a commodity. Companies are interested in comparing private and public services before deciding which service to use. Figure 1.11 summarizes private call minute costs for a number of companies. The results can be used in further investigations as to why the private phone charges become higher compared to other companies, as well as deciding whether fully or partially private services should be used for voice. It is recommended that the investigation be repeated periodically because the tariffs are subject to frequent changes.

1.10 Comparing Costs of Kilobits

Because of the competition between carriers in many countries, the costs for transmitted kilobits can be significantly different. Figure 1.12 shows the costs for a number of companies. This result does not detail the topology of the network, the suppliers, the level of redundancy, or the nature of the carried traffic. The results can aid in further investigation as to why private phone charges become higher and to investigate the costs of unexpectedly higher prices. This sort of a gap analysis can lead to a change of suppliers, topology, redundancy, protocols, or a combination of changes. It is recommended that the investigation be repeated periodically because the tariffs are subject to frequent changes.

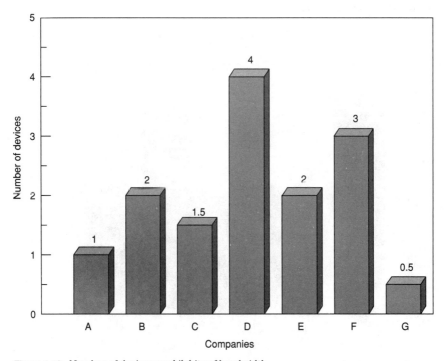

Figure 1.10 Number of devices per kilobits of bandwidth

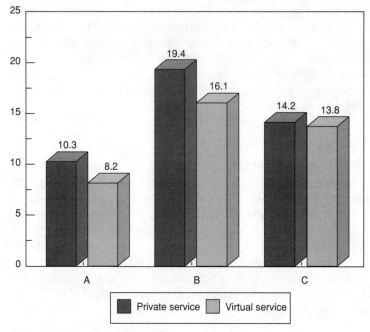

Figure 1.11 Comparing call-minute costs

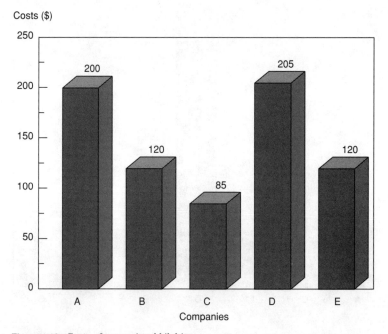

Figure 1.12 Costs of transmitted kilobits

1.11   Service Quality

Service quality can also be characterized by the distribution of problem (fault) resolution time. The trouble tickets processed and evaluated by the author and the distribution functions for two companies are shown in Figure 1.13. Company A (Figure 13a) is very typical in that the majority of problems can be resolved within a relatively short period of time. Subtle problems need more time, but there are not too many of these. Company B (Figure 1.13b) shows a peak of 3 and 4 hours, indicating administrative delays when referring trouble tickets to technical maintenance or outside vendors. These results can help cost-justify more or better tools or more people to support troubleshooting.

1.12   Outsourcing or Insourcing

The demands of benchmarking can also be viewed from another perspective. Companies that are interested in outsourcing network management might want to compare those companies that can take over the network management functions, equipment, facilities, and people. In addition, the outsourced network of Company A or B can offer new markets to the outsourcer. Important benchmarking results for this indicator are summarized in Table 1.2. The indicators used were borrowed from outsourcing diligency studies but the concrete numbers were changed by the author.

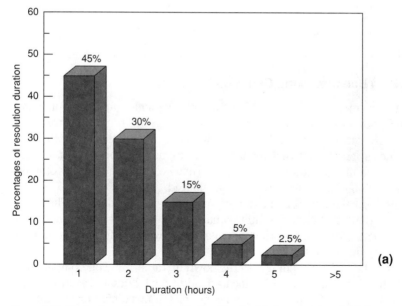

Figure 1.13   Evaluation of problem resolution time for (a) Company A and (b) Company B

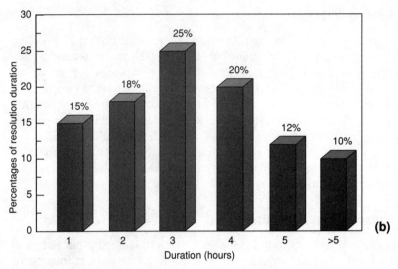

Figure 1.13 *(Continued)*

TABLE 1.2 Benchmarking as Preparation of Outsourcing Bids

Indicators	Company A	Company B
Number of staff	182	224
Global coverage	all continents	3 continents
Facility management	good	excellent
Value of equipment	$427 million	$615 million
Facilities leasing expenses (annually)	$92 million	$82 million
Operating costs (excluding facilities)	$72 million	$87 million
Average age of staff	38 years	41 years

1.13 How Benchmarking Can Highlight Gaps

To demonstrate how benchmarking can help highlight gaps, the following example shows two sets of results. One set of data was received during the first benchmark; the second set approximately 6 months later. In both cases, the support of network management functions was investigated, particularly in fault, configuration, and performance management. In total, 80 network management functions were evaluated.

Altogether, 80 network management functions were evaluated. The number of functions evaluated for fault management was 40, while 10 functions were evaluated for configuration management. Twelve functions were used for performance management.

The results of the benchmarks are shown in Table 1.3. The company invested into instruments and allocated additional human resources for benchmark 2 after evaluating the reasons for the unsatisfactory results of benchmark 1.

In reference to the total number of functions under consideration, the company could increase the full support by 25 percent, increase partial support by 50 percent, and reduce the number of unsupported functions by 33.33 percent. Figure 1.14 graphically displays the results of both benchmarks. This benchmark was conducted in a network management center that served multiple outsourcing clients.

TABLE 1.3 Results of Two Benchmarking Exercises

Function	Fully supported		Partially supported		Not supported	
	B1	B2	B1	B2	B1	B2
Fault management	10	15	20	20	10	5
Configuration management	2	3	6	6	2	1
Performance management	2	4	6	7	4	1
	•	•	•	•	•	•
	•	•	•	•	•	•
	•	•	•	•	•	•
Total	20	25	30	45	30	10

Number of requirements

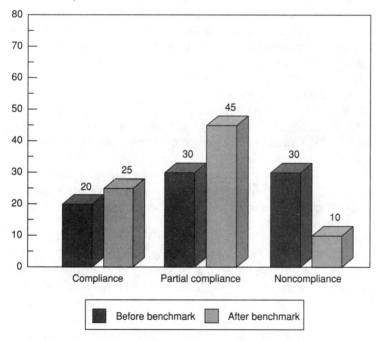

Total requirements = 80

Figure 1.14 Improvement in supporting network management functions

1.14 Summary

Network management audits and benchmarks can be used for the following purposes:

- to highlight gaps in network management processes, instrumentation, standards, and human resources
- compare results to the industry average or best practice
- to prepare outsourcing decisions

In most cases, these three objectives are combined to increase the efficiency of the benchmark.

Benchmarking is not without problems. A number of items can cause problems and impact the results of the benchmarks. For example, the most important issue is that the industry average does not contain enough samples. Thus, a comparison with the specific results of the company is not truly representative. Also, errors in measurements can occur when not using the right instruments or using incorrect settings of parameters. Dishonest or biased interview partners can also skew answers in the questionnaires and interviews, making it impossible to filter out true answers for the benchmarking team.

A lack of documentation on processes, functions, instruments, and protocols can significantly delay the data collection phase. Thus, the total lead time of benchmarking is high risk. Obsolete documentation can lead the benchmarking work in a false direction. Usually, the team is warned too late about the out-of-date status of documentation.

False interpretation of certain indicators can be caused by the specific use of certain indicators within the company. If the team is informed about this error, however, ad hoc redefinitions can easily solve this problem. The nonapplicability or nonavailability of monitors can considerably delay the data collection phase. Manually collected data are no replacement for accurately monitored data.

Finally, the company might not permit observations of operations in an "end-to-end-mode"; in particular, results can be significantly impacted when shift takeovers, the performance of the client contact point, and remote diagnostics cannot be analyzed end-to-end.

Most likely, all these negative impacts do not happen during the same benchmark, but some might. To avoid these impacts, each issue should be addressed during the very first meeting between the company and the benchmarking team.

2

Benchmarking Phases

The principal phases of network management audits and benchmarks are data collection, comparison with best practice or industry average, gap analysis, recommendations, and presentations. Figure 2.1 summarizes all the benchmarking phases. Each phase is also described in the following subsections.

2.1 Data Collection

Data collection consists of documentation, interviews, observations of efficiencies, and comparisons. Documentation is the analysis of topologies, managed objects, performance reports, network management instruments, and closed trouble tickets.

Interviews with various persons in the network management organization also take place. The interviews ask what the major problems are; what feasible solutions exist; and the job assignments, education, and skill levels of the personnel. The observations to rate operational efficiency concern the reaction to trouble calls, how management tools are used, what the support quality is for various shifts, shift-takeover processes, and documentation.

2.2 Data Consolidation

Comparisons are made with best practices or industry averages and include either the specific branch or a cross-branch evaluation. Usually, cross-branch indicators are used first. Table 2.1 displays the results of such a comparison, indicating considerable differences in certain areas between industry and client.

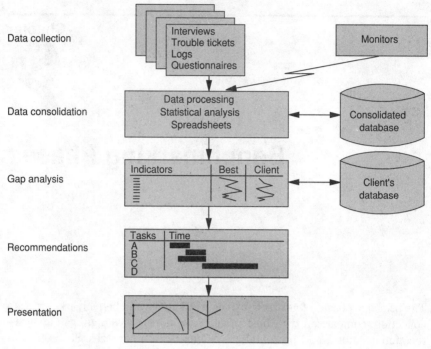

Figure 2.1 Flowchart of benchmarking phases

TABLE 2.1 Sample Comparison of Benchmarking Indicators

Benchmarking indicator	Composite average	Company ABC benchmarking result
Mean time to repair	4 hours	6 hours
Average total fault time	12 hours	36 hours
End-user workstation availability	98.50%	96.75%
Proactive versus user-reported troubles	75% proactive	25% proactive
Chronic problem tickets closing time	1 month	3 weeks
Percent of trouble calls abandoned	2% or less	10%
Average delay of answering trouble calls	15 sec. or less	30 sec.
Fault log defects	80% or higher accurate	maximal 60%
Percent new service orders completed on time	95%	68%
Mean time between failures		
Multiplexers	2 years	not measured
Routers	1.5 years	not measured

2.3 Gap Analysis

Gap analysis provides real detail and addresses the functions of network management, such as client contact point activities, operations support, fault tracking, change control, security management, design and planning, finance and billing, fault monitoring, performance monitoring, implementation and maintenance, security management, and systems administration. Instruments such as monitors, analyzers, element management systems, integrators, and modeling tools are addressed in-depth. Standards are classified as proprietary, de facto, or open. The assignment of human resources to processes and instruments is also investigated in-depth.

Evaluating the completed forms needs some interpretation depending on the networking culture of the client. To demonstrate the amount of work required, the following is a short list of network management functions and generic network management related instruments.

Network management functions:

- client contact point
- operation support
- fault tracking
- change control
- planning and design
- finance and billing
- implementation and maintenance
- security management
- fault monitoring
- performance monitoring
- systems administration

Network management instruments:

- network management integrators
- WAN element management systems
- LAN element management systems
- monitors and analyzers
- security management systems
- administration instruments
- database tools
- client control point instruments

2.4 Elaborating the Recommendations

The gaps identified in the previous phase become the basis for improvement or for supporting the decision to outsource certain network management functions. If insourcing all functions, the company must set priorities and timeframes and estimate the demand on tools, equipment, and human resources.

2.5 Benchmarking Tools

Benchmarking needs a unique set of instruments; no single instrument can cover all phases. Rather, a number of productivity tools are expected to be used in combination. The most important tools include the following:

Word processors. Most interviews are typed and prepared for distribution. The clients are expected to fill in the questionnaires and send them back to the benchmarking company. This same tool can be used during on-site interviews as well.

E-mail packages. Communication between the benchmarking company and the client should be arranged electronically. In most cases, word processing files can be embedded into E-mail packages. Faxes can be used for graphics if the E-mail package does not provide the necessary quality.

Spreadsheets. Spreadsheets should be used to maintain and compute indicators. Any type of form that requires summarizations or calculations can be entered into spreadsheets.

Data processing applications. For larger calculations, simple data processing programs can be used; either existing packages or custom developmental programs can be chosen.

Statistical packages. For special-purpose statistical analysis, statistical packages can be useful; they represent the high-end analysis in areas where spreadsheet products do not provide powerful enough solutions.

Databases or files. Databases serve a dual goal. During benchmarking, one needs workfiles for the questionnaires and measurement data if monitors are used. In addition, larger amounts of data must be maintained for the clients. These data can be used for comparative analysis or individually if companies periodically repeat auditing and benchmarking.

Query tools. These tools guarantee easy access to databases and enable users to make competitive analyses across multiple clients. Query tools are sometimes combined with reporting software.

Presentation tools. These tools permit the visualization of existing results, making the interpretation much easier. Most of these tools take advantage of the graphics capabilities of workstations and printers. The customization demand, however, should not be underestimated.

Report generators. Report generators represent a special class of application software with a large number of default reports and only require that parameters be defined. These packages can substantially increase the efficiency of reporting.

Monitors. These tools are occasionally used for information collection. They usually run in LAN segments as probes or in servers as software monitors. Monitors provide a snapshot of service quality and resource utilization.

Modeling packages. To support performance prediction after recommended improvements, modeling tools (using simulation and applied-queuing) are used. Modeling can cost-justify the recommended improvements.

CD-ROM. CD-ROM, or compact disc read-only memory, is an alternative method to store data, distribute questions, and receive answers. The applicability depends on the client.

Project management software. Both the benchmarking phases and the recommended activities should be planned and properly supervised. PC-based project management packages are very helpful and are inexpensive and easy to maintain and install.

2.6 Benchmarking Personnel

Benchmarking network management is not a trivial exercise. Usually, small teams of two to five people are assigned to benchmarking work because teamwork is necessary. Two different types of persons conduct benchmarks: senior consultants and analysts. Skill levels overlap in certain areas, but usually both types of professionals have specific areas of expertise. Table 2.2 summarizes the required skills for the benchmarking team members. The required skill levels are very ambitious; it is highly unlikely that junior consultants would lead a benchmarking project.

TABLE 2.2 Skill Requirements for Benchmarking

Skills	Senior consultant	Analyst
General understanding of communication technologies	x	
WAN components (modems, muxes, switches, protocols, signaling, topologies, network operating systems)		x
LAN components (bridges, routers, switches, protocols, signaling, topologies, network operating systems)		x
MAN components (nodes, repeaters, switches, protocols, signaling, topologies, network operating systems)		x
General understanding of network management functions	x	
Configuration management		x
Fault management		x

TABLE 2.2 Skill Requirements for Benchmarking (Continued)

Skills	Senior consultant	Analyst
Performance management		x
Security management		x
Accounting management		x
General understanding of network management instruments	x	
Integrators	x	x
Element management systems	x	x
Monitors and analyzers		x
Modeling devices and simulators		x
Administration tools	x	x
General understanding of network management platforms	x	
Platform attributes		x
Leading products, such as OneVision, SunNet Manager, OpenView, AIX NetView	x	x
General understanding of responsibilities and qualifying experiences of human resources	x	x
Voice networks	x	
Data networks	x	
Mobile networks	x	
Video networks	x	
Combination of various networks	x	
Basic understanding of costing and chargeback policies and technologies		
Costing for hardware, software, communications, people, and infrastructure	x	x
Chargeback techniques	x	
Good analytical capabilities		
Ad-hoc correlations	x	
Improvisation with interview questions	x	
Diligence in collecting and processing data		
Completeness of collected information	x	x
Checker functions for redundancy and credibility of input data	x	x
Communication skills	x	
With top management	x	
With management	x	x
With technicians		x
Ability to work out concepts		
Hypothesis	x	
Feasibility checks	x	

Skills	Senior consultant	Analyst
Basic knowledge of statistics		
Mean value	x	x
Distribution functions	x	x
Percentiles	x	x
Density functions	x	x
Experience with statistical packages	x	x
Project management		
Resources allocation	x	
Milestone control	x	
Delivery control	x	
Ability to pursue work at detail level		
Evaluating logs		x
Monitoring		x
Conducting special measurements		x
Awareness of state-of-the-art network management strategies and concepts		
Client server	x	x
Management applications	x	x
Distributed management	x	x
Object-oriented databases	x	
Integration	x	x
Automation	x	x

2.7 Financial Resources

Benchmarking also needs financial resources. The price depends on the number of the following network management functional areas:

- managed objects to be benchmarked
- network management centers to be visited
- managers to be interviewed
- subject-matter experts to be interviewed
- network management integrator products in use
- network element management systems in use
- monitors in use
- network management protocols in use
- communication forms, such as data, voice, image, and video to be managed

The pricing algorithm is

$$
\begin{aligned}
\text{Price} = c0 + (((((&(c1 \times \text{no. of managed objects}) + (c31 \times \text{no. of managers}) \\
&+ (c32 \times \text{no. of SMEs}) + (c4 \times \text{\# integrators}) + (c4 \times 0.5 \times \text{EMSs})) \\
&+ (c4 \times 0.2 \times \text{no. of monitors})) \times (1 + c5)) \times (1 + c2)) \times (1 + c6)) \\
&\times (1 + c7))
\end{aligned}
$$

where:

    c_0 = $5000 for covering initial project costs

    c_1 = $5 for 1 to 4,999 managed objects
         $3 for 5000 to 24,999
         $1 for over 25,000

    c_2 = 0 weighting factor for 1 site management center
         0.1 for 2
         0.2 for 3
         0.3 for 4 or higher

   c_{31} = $50 cost for interviewing managers

   c_{32} = $30 cost for interviewing subject-matter experts (SMEs)

    c_4 = $500 evaluation cost for management instruments

    c_5 = 0 weighting factor for 1 network management protocol
         0.1 for 2
         0.2 for 3 or higher

    c_6 = 0 weighting factor for 1 communication form (voice, data, image, or video) supported by benchmark
         0.1 for 2
         0.2 for 3 or higher

    c_7 = 0 weighting factor for 1 management functional group evaluated during benchmark
         0.3 for 2
         0.4 for 3
         0.5 for 4 or higher

  EMS = element management system

Financial examples

Two practical examples are shown in Table 2.3; one for a relatively small network and one for a larger network. The parameter settings are also shown. Using the pricing algorithm, the results are shown in Table 2.4.

These two examples show the range of costs charged by the benchmarking company for conducting the network management benchmark. Additional expenses on behalf of the company under benchmark can include other items, which are described in the following subsections.

Human resources. The assignment of personnel to the benchmarking team can represent a substantial amount of money when the time to conduct the benchmark is long. Personnel are usually taken away from other projects, causing project delay or overtime elsewhere. It is recommended to consider the unloaded human resource costs corresponding with the annual salary divided by 1600 hours. With $100,000 as a high-end salary average, the support hour would cost $62.50 for a subject-matter expert.

TABLE 2.3 Parameter Settings for Two Networks

Parameters	Case 1	Case 2
Managed objects to be benchmarked	1,000	30,000
Network management centers to be visited	1	5
Managers to be interviewed	3	20
Subject matter experts to be interviewed	15	110
Network management integrator products in use	0	3
Network element management systems in use	3	21
Monitors in use	2	18
Network management protocols in use	2	5
Communication forms, such as data, voice, image, and video, to be managed	1	2
Network management functional areas included in the benchmark	3	5

TABLE 2.4 Results of Pricing Algorithm

	Basic price	Basic price + 25% profit
Case 1	$16,004.00	$20,005.00
Case 2	$116,295.90	$145,369.87

Instrumentation. Existing instruments and tools are usually used by the benchmarking team. Occasionally, additional equipment can be rented or leased for the duration of the project to monitor certain indicators and process measurement data. The rough estimate for these expenses is between $1000 and $5000.

Project administration. This expense item is often overlooked and ignored. Compared to the benchmarking costs invoiced by the benchmarking company and human resources costs, it is not significant. Project administration includes typing, word processors, and graphical software. These tools are available either at the company premises or at the site of the benchmarking company. If necessary, this expense source can be itemized for between $500 and $2000.

Saving costs

To help cost-justify benchmarking projects, it is beneficial to define all achievable improvement and savings monetarily. This task is not always easy. To use financial analysis techniques, the items shown in Table 2.5 should be quantified. Minimal and maximal saving estimates are provided for both cases, but the timeframe is not quantified, because it depends on the actual action plan of the company. Average timeframes are between 3 and 18 months if action plans are realistic and their implementation has received top management support. All saving components are listed in Table 2.5, but, in certain cases, data were not available for the payback analysis.

TABLE 2.5 Cost Savings Seen by Benchmarking

Saving items	Case 1		Case 2	
	Optimistic	Pessimistic	Optimistic	Pessimistic
Saving components in network management functions				
Downtime reduction by better fault tracking*				
Increase of number of faults solved by one referral*				
More widely used proactive monitoring*	50,000	25,000	100,000	50,000
Faster change management process	10,000	10,000	40,000	40,000
Elimination of paper forms	-----	-----	-----	-----
Automation of routine process steps	-----	-----	-----	-----
In-time capacity upgrades	5,000	5,000	10,000	10,000
Less violations of service-level agreements, resulting in less penalties	10,000	10,000	20,000	20,000
Subtotal	75,000	50,000	170,000	120,000
Saving components in network management instrumentation				
Avoidance of redundant instruments	12,000	6,000	18,000	12,000
Provide instrumentation to most or all functions	-----	-----	-----	-----
Support by console management eliminating consoles	8,000	4,000	14,000	8,000
Consolidation of help desks Spend once for integrated management	-----	-----	-----	-----
Subtotal	20,000	10,000	32,000	20,000
Saving components in human resources				
In-time education	-----	-----	-----	-----
Hiring for missing jobs and positions	-----	-----	-----	-----
Highlighting where labor hours are spent	-----	-----	-----	-----
Avoidance of idle time of staff	12,500	7,500	25,000	15,000
Subtotal	12,500	7,500	25,000	15,000
Total	107,500	76,000	227,000	155,000

*These items contribute to increase the overall availability of the network—in these cases, between 1 and 2%.

Financial analysis techniques

To evaluate the financial feasibility of network management benchmarks, three financial analysis techniques can be used for the business: payback analysis, cash flow analysis, and return on investment (ROI) using present-value analysis.

As a result of conducting many network management benchmarks, clients usually cost-justify network management benchmarks using payback analysis. In other areas of network management, such as systems integration, purchasing an integrator, or reengineering network management processes, all three techniques are used.

Payback analysis

Payback analysis determines when the network management benchmark can pay for itself. The payback period can be computed as follows:

$$\text{Payback period} = \frac{\text{Benchmarking costs (one-time and periodic)}}{\text{Net savings achieved by benchmarking annually}}$$

The equation can be used for the two cases shown in Table 2.5 and 2.6.

To calculate net savings, the 2 years after conducting the network management benchmarks are aggregated. The payback period is shown in Table 2.7.

TABLE 2.6 Optimistic versus Pessimistic Cost Savings

	Case 1		Case 2	
	Optimistic	Pessimistic	Optimistic	Pessimistic
Cost components				
Benchmarking	$20,005	20,005	145,369	145,369
Human resources	2,000	5,000	10,000	15,000
Instrumentation	1,000	2,000	2,000	5,000
Administration	500	1,000	1,000	2,000
Total	23,505	28,005	158,369	167,369
Savings components				
Processes	75,000	50,000	170,000	120,000
Instruments	20,000	10,000	32,000	20,000
Human resources	12,500	7,500	25,000	15,000
Total	107,500	67,000	227,000	155,000

TABLE 2.7 Payback Period Calculations

Case 1	Optimistic	0.27 years	=	2.6 months
	Pessimistic	0.42 years	=	5 months
Case 2	Optimistic	0.69 years	=	8.3 months
	Pessimistic	0.93 years	=	11.2 months

The examples in Tables 2.5 and 2.6 demonstrate how to compute the payback period. The numbers might be completely different for other companies.

2.8 Summary

Each company must assess costs and savings before reaching a decision for a network management benchmark. Usually, building the business case is short, and top management—who basically supports the idea of benchmarking—is generous for deciding for it. They also consider nonquantifiable benefits such as the knowledge of the position of the company against peers and the industry average or the use of benchmarking results to build a case for larger investments.

Critical Network Management Success Factors

3.1 Introduction

Prior to auditing or benchmarking, the *critical success factors* of network management need to be addressed. Critical success factors are those key areas of activity in which favorable results are absolutely necessary for the network management organization to reach its goals. Critical success factors for network management are the following:

Processes: the sequence of application steps, including guidelines for how to use tools to execute network management functions.

Instruments: hardware and software, or both, for collecting, compressing, databasing, and correlating network management information and predicting the future performance and use of network components.

Standards: agreements for how to store, process, and transmit network management related information.

Human resources: all individuals involved in supporting network management functions, relating to the organization of the company and depending on the degree that systems and networking resources are shared as a function of the various business objectives.

To quantify the quality of network management, indicators must be defined for each of the critical success factors.

3.2 Business Model of Network Management

Before discussing principal management functions, the business model of
the processes is shown. The logical model mirrors the usual distribution of
management functions among different departmental organizations. To
avoid duplications, management functions are allocated to a single business
area only. Assuming shared management knowledge, information integrity
can be guaranteed between business areas.

Figure 3.1 shows the logical model highlighting business areas and the major information flows between them. This model is based on the recommendations of the Network Management Forum. Terplan (1994) has adopted
and extended this model by customizing network management functions
and processes to practical needs. The scope of business areas is based on an
average corporation, but it is not the model of any particular corporation.

The entry point into the model is the client. Clients represent internal or
external customers or any other users of management services. Clients can
report problems, request changes, order equipment or facilities, or just request information. This interface is the *client contact point*, implemented
as the single point of contact to handle all client-related problems, changes,
orders, and inquiries. Principal activities of the client contact point include
the following:

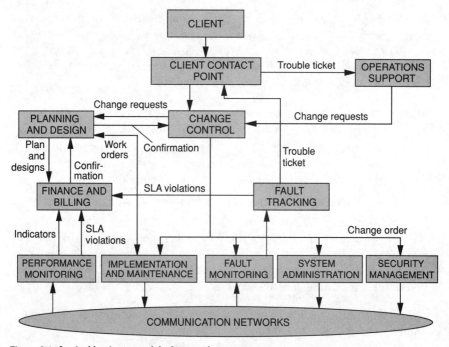

Figure 3.1 Logical business model of network management

- receiving problem reports
- handling calls
- handling enquiries
- receiving change requests
- handling orders
- making service requests
- opening and referring trouble tickets
- closing trouble tickets

Operations support then receives trouble tickets from the client contact point. Major activities of operations support include the following:

- determining problems using trouble tickets
- diagnosing problems
- taking corrective actions
- repairing and replacing
- referencing to third parties
- backing up and reconfiguring
- recovering
- logging events and performing corrective actions

This business area can be further subdivided into second- and third-level support, vendor support, and third-party maintainers.

As a result of troubleshooting, change requests for fixes are sent to *change control*. Not only clients, but computerized monitors can report problems to operations support. In this case, trouble tickets are opened in the *fault monitoring* department and forwarded to operations support via the *fault tracking* department. Fault tracking concentrates on the following:

- tracking manually reported or monitored faults
- tracking the progress and escalation of problems if necessary
- distributing information
- referring problems

Fault tracking is a very central activity that plays a key role in supervising and correcting service quality related problems.

Change control deals with managing, processing, and tracking service orders; routing service orders; and supervising the handling of changes. The results of these activities are validated change requests sent to the *plan-*

ning and design department. This business area supports design and planning functions, such as the following:

- analyzing needs
- projecting application load
- sizing resources
- authorizing and tracking changes
- raising purchase orders
- producing implementation plans
- establishing company standards
- maintaining quality assurance

The results of planning and design are distributed to *finance and billing* and *implementation and maintenance*. Implementation and maintenance implements changes and work orders sent by planning and design and change control. In addition, this area is in charge of the following activities:

- maintaining resources
- inspection
- maintaining the configuration database
- provisioning

By continuously monitoring systems and networks, status and performance information can be collected. Fault monitoring proactively detects problems and opens and refers trouble tickets.

Performance monitoring is concerned with the following issues:

- monitoring system and network performance
- monitoring service-level agreements (SLAs)
- monitoring third-party and vendor performance
- optimizing, modeling, and tuning
- reporting on usage statistics and trends to management and users

Performance monitoring informs finance and billing about service quality. Fault monitoring opens trouble tickets and sends them to fault tracking.

Security management is responsible for ensuring secure communications and protecting the management system. The following functions are supported:

- threat analysis
- administration (access control, partitioning, authentication)
- detection (evaluating services and solutions)
- recovery (evaluating services and solutions)
- protection of management systems

Systems administration is a key activity that requires the coordination of many aspects and is responsible for the administration of the whole distributed processing environment, including the following:

- software version control
- software distribution
- systems management (upgrades, disk space management, job control)
- administration of user-definable tables (user profiles, router tables, security servers)
- local and remote configuration of resources
- management of names and addresses
- applications management

Finance and billing is the focal point for receiving status indicators, SLA violations, plans, designs, changes, and invoices from third parties. Finance and billing is concerned with the following as well:

- asset management
- costing services
- client billing
- usage and outage collection
- calculation of rebates to clients
- bill verification
- software license control

All business areas are supported by instruments with different levels of sophistication. Section 3.3 addresses the principal instruments.

Form 3.1 summarizes all the network management functions. In completing the form, users answer the following questions:

- What are fully supported functions?
- What are partially supported functions?
- What functions are not supported?
- In case of full or partial support, who is the responsible party?

FORM 3.1 Network Management Functions

	Fully supported	Partially supported	Not supported	Responsible party
Client Contact Point				
Receiving problem reports				
Handling calls				
Handling enquiries				
Receiving change requests				
Handling orders				
Making service requests				
Opening and referring trouble tickets				
Closing trouble tickets				
Operations Support				
Determining problems using trouble tickets				
Diagnosing problems				
Taking corrective actions				
Repairing and replacing				
Referencing to third parties				
Backing up and reconfiguration				
Recovering				
Logging events and performing corrective actions				
Fault Tracking				
Tracking manually reported or monitored faults				
Tracking the progress and escalation of problems if necessary				
Distributing information				
Referring problems				
Change Control				
Managing, processing, and tracking of service orders				
Routing service orders				
Supervising the handling of changes				
Planning and Design				
Analyzing needs				
Projecting application load				
Sizing resources				
Authorizing and tracking changes				
Raising purchase orders				
Producing implementation plans				
Establishing company standards				
Maintaining quality assurance				

	Fully supported	Partially supported	Not supported	Responsible party
Finance and Billing				
Asset management				
Costing services				
Client billing				
Usage and outage collection				
Calculation of rebates to clients				
Bill verification				
Software license control				
Implementation and Maintenance				
Implementing change requests and work orders				
Maintaining resources				
Inspection				
Maintaining the configuration database				
Provisioning				
Fault Monitoring				
Monitoring system and network for proactive problem detection				
Opening additional trouble tickets				
Referring trouble tickets				
Performance Monitoring				
Monitoring system and network performance				
Monitoring service-level agreements				
Monitoring third-party and vendor performance				
Optimizing, modeling, and tuning				
Reporting on usage statistics and trends to management and users				
Security Management				
Threat analysis				
Administration				
Detection				
Recovery				
Protecting the management systems				
Systems Administration				
Software version control				
Software distribution				
Systems management				

FORM 3.1 Network Management Functions (Continued)

	Fully supported	Partially supported	Not supported	Responsible party
Administration of user-definable tables				
Local and remote configuration of resources				
Management of names and addresses				
Applications management				

To help differentiate fully and partially supported functions, the following examples can be used.

Determining the problem

- Fully supported: Automatic detection using continuous monitoring of the network for key problem indicators. Use of various tools for combination to identify the problem. In addition, checklists can be used and preprogrammed procedures invoked.

- Partially supported: Semiautomatic or manual detection using sampling of the network, rotating measurements, or waiting for users calling the client contact point. Few tools in combination are used to identify the problem. In addition, checklists or manual procedures can be used.

Supervising the handling of changes

- Fully supported: A change management tool that is linked to the inventory or asset database with built-in updating and synchronization features can be used.

- Partially supported: Change management forms that are manually entered into files and databases can be used.

Resource sizing

- Fully supported: Performance databases, measurement devices, and modeling packages to accurately determine the resource demand of key facilities and equipment that can be used.

- Partially supported: Performance data and unsophisticated—mostly spreadsheet based—instruments to determine the resource demand of key facilities and equipment that can be used.

Protecting management systems

- Fully supported: Providing additional security and protection services, e.g., chip keys, chip cards, or biometrics or Kerberos-based authentication solutions.

- Partially supported: Taking full advantage of the abilities of the operating system of the network management product.

Optimizing, modeling, and tuning

- Fully supported: Proactive performance optimization by a small team using a combination of monitors, analyzers, and modeling packages.
- Partially supported: Reactive performance optimization by a small team using few tools.

Chapters 5, 6, and 7 discuss the practical results for benchmarking network management.

3.3 Principal Network Management Processes

Network management functions are the basis for the processes. Depending on the process under consideration, the number of functions included can vary greatly. Substantial efficiency improvements can be expected when reengineering network management processes. *Reengineering* requires a focus on the following items:

- investigating the existing process flow
- identifying manual and semi-manual steps
- quantifying time spent on process steps
- determining wait times and interfunctional relationships
- identifying media changes during process execution
- confirming information sharing between functions and processes
- determining whether functions can be further decomposed
- checking which manual steps can be automated and how

Companies have reported surprising results after reengineering. One of the expected results from benchmarks is an answer to the process-quality questions. This segment focuses on only two processes: change management and fault management. In both cases, two practical alternatives are presented (Terplan 1994). Other important processes are referenced by indicating typical instruments in use.

Despite the fact that processes are used extensively, it is extremely difficult to design, develop, and implement standard processes applicable to companies across industries.

Configuration, asset, and inventory data are necessary to support network management functions and processes. Currently, benchmarking can identify the ownership of multiple databases and files within various organizational units. Many organizations and standard bodies want to unify activities for an enterprise-wide repository on the basis of object-orientation. This goal, however, is a long way off. Only fragmented solutions exist today that use relational technology.

The Network Management Forum has compiled a very valuable list of attributes for the principal managed objects. Table 3.1 shows examples of equipment and circuits from this list. In most cases, the integrated database is not the only network management related database in the corporation. Typical databases support the network management product platform, trouble management, infrastructure management, legacy management, performance management, and accounting. The integrated database is expected to be implemented with the platform products. Despite high-level integration, special data remain in the databases of element management systems (EMS), monitors, and network management applications. In all cases, communication between the databases is necessary; for exporting and importing data, structured query language (SQL) seems to be the de facto industry standard.

Today, database implementations are based on relational technology; the future is, however, object orientation. This technology has a lot of potential, with savings of up to 70 percent for storage requirements.

TABLE 3.1 Samples for Attributes of Managed Objects

M-Object-Class	Mandatory attributes	Optional attributes
Circuit	administrative State aEndPointName circuitBandwidth circuitID circuitType componentNames objectClass operationalState transmissionDirection zEndPointName	circuitGroupName CompositeCircuitNames contactNames customerNames facilityNames networkNames providerNames serviceNames typeText userLabels
Equipment-Bridge	administrativeState contactNames equipmentID locationName objectClass operationsState packetLossRate packetLossRateThreshold vendorName	customerNames equipmentType functionNames manufacturerName manufacturerText networkNames productLabels release serialNumber serviceNames softwareNames typeText userLabels

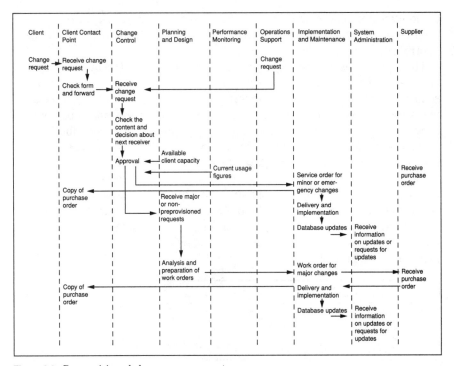

Figure 3.2 Preprovisioned change management process

As shown in Figure 3.1 (Terplan, 1994), change control should be the focal point of change management. Basically, there are two different types of change requests or service orders: *preprovisioned requests* and *non-preprovisioned* requests. Preprovisioned requests deal with minor changes. These can be accepted by the client contact point, or help desk, without checking with other management groups. Non-preprovisioned requests deal with major changes that require a coordinated approach between various management groups.

Preprovisioned change management process. The preprovisioned change management process is shown in Figure 3.2. Typically, these change requests arrive from one of two sources: the client contact point or the operations support group.

In the first case, clients (users) send their service orders in the form of change requests to the client contact point. Changes can take the form of scheduled network, systems or software upgrades, ongoing network and systems evolution, moves, additions, and the resetting of certain parameters. Depending on the nature of the change, the planning and design group could be involved, although including this group can be too time-consuming to be effective. The client contact point receives the requests, evaluates their completeness, classifies them, and then forwards them to the change control group.

The operations support group sends other types of requests to the change control group; these are in most cases fixes or emergency changes required to resume operations. In any case, the change control group evaluates the content of the requests before approval for execution or before forwarding them to the planning and design group for further evaluation. It is recommended that standard change request forms be used that contain the information identified in Figure. 3.3.

Usually, change request forms are provided and updated electronically. If the change control group is dissatisfied with the completeness of the form, requests must be sent back to the originators for improvement. End users must typically undergo an educational process for using change request forms; after the second or third refusal, persons generating change requests learn how to fill out the forms.

Change Coordinator Segment
Identification of change Change number Date of request Identification of change coordinator
Change Requester Segment

Requester name, affiliation, phone, fax, E-mail
Location
Change description
Network components involved by inventory/asset identification
Network components affected by change
  Minor impact = change is nondisruptive
  Regular impact = change might be disruptive
  Major impact = change is disruptive
Due date
Priority
Reason for change
Estimated duration of change
Personnel involved in executing change
Fallback procedure when change fails

Approval Segment
Date of approval Signature
Evaluation Segment
Result of change Downtime due to change Cancellation or postponement Database updates after successful change

Figure 3.3 Sample information on change request form

Occasional consultation is necessary with the planning and design group to ensure that the client still has adequate capacity, e.g., bandwidth or ports in multiplexers or routers to handle the change. The performance monitoring group can also be asked to provide usage statistics on the resources involved in the change. With minor changes, however, these dialogs are usually unnecessary.

The approval form contains an accurate schedule and designated responsibilities for other management groups. Frequently, single changes invoke a chain of additional change requests. The change control group might need to use automated project management tools to supervise changes. Approval is granted for minor or emergency changes to the implementation and maintenance group. In the case of major changes, the planning and design group must issue work orders. In both cases, but, with different urgency, purchase orders are issued to the suppliers. After delivery of parts and components, changes can be implemented, followed by tests and the necessary database updates. Documentation of the change control process includes the following:

- approval reports
- change summaries
- components affected by changes
- implementation schedules

Finally, the system administration group and the client contact point are provided with the documentation of executed purchase orders and changes.

The non-preprovisioned change management process. Figure 3.4 shows the flow of non-preprovisioned change requests or service orders. The provisioning of such orders requires a complete specification to design and implement the modifications necessary to the existing network. The client contact point forwards the order after logging it to the change control group. Change requests are assigned when the content of the request is complete. The same form can be used as with preprovisioned change orders. The planning and design group is much more involved here than with preprovisioned change requests. Using an accurate representation of the current inventory, configuration and spare capacity on facilities and equipment, the modifications and extensions can be designed. The design must be evaluated in terms of technological and financial feasibility. The finance and billing group must also issue its approval. Upon receiving approval, detailed design plans are generated. Modeling tools can also be used to model disaster scenarios and validate the request originated by the change control group.

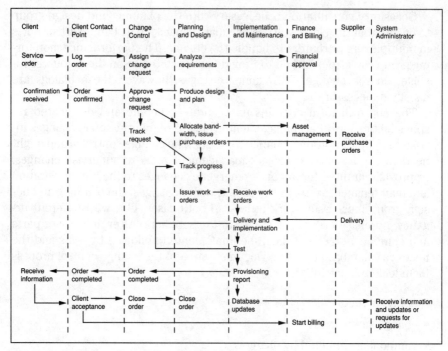

Figure 3.4 Non-preprovisioned change management process

The planning and design group then produces a detailed implementation plan that describes all activities that must be completed to effect the order. It defines work order packages, dependencies, and time scales. After approval, the client control point confirms the order to the client.

The change control group tracks the execution of the implementation plan; the planning and design group must immediately report any deadline violations to the change control group. After receiving the approved change requests, the planning and design group allocates equipment and bandwidth and informs finance and billing about ordered assets. The purchase orders are sent to the suppliers, and progress is tracked by the planning and design group.

After the implementation and maintenance group carries out work orders, it begins testing. The group must report all major milestones to the planning and design group. On the other hand, the change control group receives supervisory information from the planning and design group.

After printing the final provisioning report, an "order completed" message is sent to the client. After approval, the service order is closed, inventory and configuration databases updated, and billing initiated.

Typical instruments that support these change management processes are listed in Table 3.2.

TABLE 3.2 Typical Instruments for Change Management

Network management instruments	Primary use	Secondary use
Integrators		
Manager of managers	x	
Management platforms	x	
Element Management Systems for Wide Area Networks (WANs)		
Modems	x	
Multiplexers		
Packet switches		
Fast packet switches		
ATM switches		
SMDS nodes		
ISDN nodes		
Mobile communication		
Matrix switches		
Operations support systems		
Element Management Systems for Local Area Networks (LANs)		
Bridges	x	
Routers		
Brouters		
Repeaters		
Extenders		
Hubs		
Segments		
FDDI		
DQDB		
Monitors and Analyzers		
WAN monitor		
LAN monitor		
WAN analyzer		
LAN analyzer		
Network monitor		
Software monitor		
Security Management Systems		
Protection of systems and networks		x
Protection of management systems		x
Administration Instruments		
Documentation systems	x	
Modeling instruments		
Presentation tools		x
Report generators		x
Trouble-tracking tools		x
Software distribution tools		
Software licensing tools		
Database Tools		
Databases	x	
MIB browser	x	
Inquiry tools	x	
Client Service Point Instruments		
Prediagnosis by phone		
Automated call distributor		
Voice mail		
E-mail		
Pager		
Console emulator		
Expert system		

There are two aspects of fault handling (Terplan, 1994): faults reported by users (client-reported faults), and faults detected by network monitoring or management instruments. In either case, many of the functional steps are the same, as shown in Figures 3.5 and 3.6.

Client-reported faults. The client typically reports a problem to the client contact point by telephone, fax, or E-mail. The client is automatically identified, and the appropriate site details and recent faults associated with the site are displayed to the client contact point. If no prior associated faults on this problem have been received, a new problem report is automatically created. Otherwise, the previous problem report is opened, assuming that management personnel were notified when problems affecting service were first reported. In other words, a new trouble ticket is opened only when it has been determined that the call is reporting a new fault.

For a new fault, the client provides fault details to diagnose the fault and determine appropriate action. Support text should be provided to assist the client control point or other first-level support personnel to help diagnose

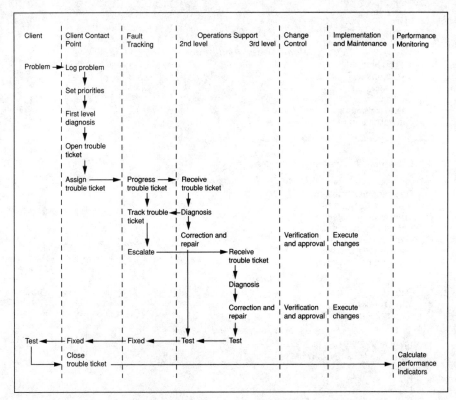

Figure 3.5 Process of client-reported faults

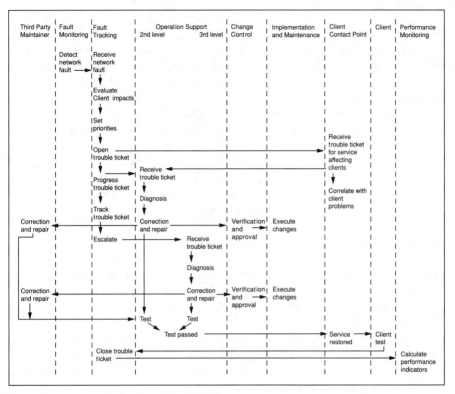

Figure 3.6   Process of instrument-detected faults

the fault and suggest possible causes. The management personnel might want to query the trouble-ticketing database for similar problems.

All dialogue between client and management personnel is logged to maintain a complete fault history that should be included into the trouble ticket when passed to operations support. Using state-of-the-art techniques, drawing and voice messages can be included as well. To enable the client contact point to assign the problem to the appropriate personnel, supporting information, such as contacts, escalation procedures, equipment, and facility attributes, are absolutely necessary.

When a problem is reported, the client control point should be able to interpret, analyze, diagnose, and correlate the problem report with other client- or network-reported problems by service reference number. After service has been restored by other management groups, the client control point must notify the client so that the client can test the service and confirm that the problem has, in fact, been resolved. If the fault is traced to equipment maintained by third-party maintenance organizations, the trouble ticket should be sent to the third party. The progress of the resolution process is then tracked by the fault tracking group. The trouble ticket should be referred to the new owner according to problem type, time of day, severity, and maintainer of the resource. If more than one action is to be taken by different parties, the trouble ticket should be sent to both. Any significant action on the trouble ticket, such as changing the ownership, must be indicated to the client. Expected time to repair should also be given to the client.

The fault-tracking group forwards the trouble ticket to the new owner, including third-party maintenance organizations, for action. In certain cases, trouble ticket information must be split into smaller units to generate actions by different owners. If possible, related trouble tickets should be associated, collapsed, and sent together to the new owner. Progress is monitored against critical milestones or time scales defined by service-level agreements. Escalation remainders notify management personnel about deadline violations. Shared fault knowledge is extremely important for anybody taking actions on a problem; in other words, the whole fault history is expected to be available to the old and new owners of the fault. Significant events must be reported back to the clients. Service-level agreements include escalation guidelines that define the time the problem should stay in any one state. A problem can remain in a pending state without escalation when progress depends on another event.

The operations support group provides both second- and third-level support for problems. It handles those calls from clients that require a technical expert to help identify the course of action. Usually, the complex problems are handled by the second-level support, and the very sophisticated, subtle problems are handled by the third-level. Third-level might also include actions for remote sites. The operations support group must have the complete fault history available, as well as details of the site and location where the fault occurred. Dialogue with the client should be kept to a minimum.

Problem diagnosis includes a number of testing procedures that might need to be invoked remotely. Both disruptive and nondisruptive tests must be considered. It is also necessary to be able to turn on performance monitoring functions for a specific period on facilities and equipment. Diagnostic tools differ greatly between wide area networks (WANs) and local area networks (LANs). Second- and third-level support personnel might need to activate predefined diagnostic and testing procedures to determine or verify where the faults are. Support personnel must be able to select the right problem-isolation procedure from a set of already defined procedures. Support personnel might want to use other network information, such as test and measurement data, to more quickly isolate faults and identify possible causes. State-of-the-art trouble tickets allow support personnel to enter the results of diagnostic tests and analysis right on the ticket. Support personnel might want to execute testing procedures online or offline. They might also want additional support data, such as dumps, status information, statistics, etc.

Management personnel can also execute corrective actions and repair. They can change and reset systems parameters, attribute values, deactivate, activate, or restart components. Management might even reconfigure parts or the whole network as corrective actions in response to faults. Change requests—excluding minor fixes—are communicated to change control for verification and approval. The actual execution of changes and repairs is performed by the implementation and maintenance group. In any case, the operations support group must ensure that the problems have been resolved via tests. If so, the fault tracking group notifies the client control point to advise clients about starting client-initiated tests. If tests are successful, and no new problems are introduced, trouble tickets can be closed and stored in the trouble-ticketing database.

Performance calculations can be completed either in near real-time or batch mode.

Faults detected by instruments.   The second alternative to client-identified faults is to use advanced fault-monitoring features within the system to detect problems in the network. If a fault occurs in the network or its systems, redundancy or fallback mechanisms designed into the network or in its systems can be automatically invoked to ensure that the effect on services is minimized. This automated correction can be either temporary or permanent. Some interruptions of service or degradation of performance, however, might occur despite the fallback. Outages and degradation should be recorded for performance evaluation and reporting. The fault tracking group receives the fault, and, to set priorities, evaluates the client impact of the fault. After setting priorities, trouble tickets are opened or extended on the basis of the nature of the fault.

When faults are detected automatically via software, fault detection includes the results of first-level diagnosis. As part of the notification, trouble tickets are sent to the client contact point to be correlated with similar or other client problems. Trouble tickets are also referred to operations support with complete fault history. If the fault is identified to equipment maintained by a third party, the trouble ticket should be sent directly to the maintainers, and the progress of the resolution process tracked by the fault tracking group. The trouble ticket should be referred to the new owner according to problem type, time of day, severity, and maintainer of the resource. If more than one action is to be taken by different parties, the trouble ticket should be sent to both. Any significant action on the trouble ticket, such as changing ownership, must be identified to the client. Expected time to repair should also be given to the client.

Handling faults is greatly assisted by self-healing and fault-tolerant technologies. Today, these technologies are incorporated into principal network equipment, such as hubs. Once faults are detected, the built-in application tries to determine the next meaningful step. This step could be the switchover to a built-in redundant component while diagnosing the problem in-depth, or starting the diagnosis and restoration functions without switchover. In most cases, the built-in intelligence is also responsible for reporting on the fault and progress to network operations.

The finance and billing group might be involved in this process by updating the asset list and occasionally calculating the refund to clients because of service outages.

Typical instruments that support fault handling processes are identified in Table 3.3.

TABLE 3.3 Typical Instruments for Fault Handling

Network management instruments	Primary use	Secondary use
Integrators		
Manager of managers	x	
Management platforms	x	
Element Management Systems for Wide Area Networks (WANs)		
Modems	x	
Multiplexers	x	
Packet switches	x	
Fast packet switches	x	
ATM switches	x	
SMDS nodes	x	
ISDN nodes	x	
Mobile communication	x	
Matrix switches	x	
Operations support systems	x	
Element Management Systems for Local Area Networks (LANs)		
Bridges	x	
Routers	x	
Brouters	x	
Repeaters	x	
Extenders	x	
Hubs	x	
Segments	x	
FDDI	x	
DQDB	x	
Monitors and Analyzers		
WAN monitor	x	
LAN monitor	x	
WAN analyzer	x	
LAN analyzer	x	
Network monitor	x	
Software monitor	x	

Network management instruments	Primary use	Secondary use
Security Management Systems		
Protection of systems and networks		
Protection of management systems		
Administration Instruments		
Documentation systems		x
Modeling instruments		
Presentation tools	x	
Report generators		x
Trouble-tracking tools	x	
Software distribution tools		
Software licensing tools		
Database Tools		
Databases	x	
MIB browser		
Inquiry tools		
Client Service Point Instruments		
Prediagnosis by phone	x	
Automated call distributor	x	
Voice mail	x	
E-mail	x	
Pager	x	
Console emulator		x
Expert system	x	

Performance management involves optimizing existing resources and no-tifying the planning and design group if existing resources are exhausted and tuning is no longer applicable. Driving forces for optimization include the following:

- data generated during continuous monitoring of the whole network
- end-user complaints about performance
- violation of service-level agreements based on evaluation of measured results
- requests from the planning and design group for performance assessment

More details can be found in (Terplan, 1994). Table 3.4 lists typical in-struments that support the process for network performance optimization.

Companies must establish policies and guidelines for the use of computer and telecommunication resources and for safeguarding information while stored or processed by a system. The policy must also address misuse or

TABLE 3.4 Typical Instruments for Performance Optimization

Network management instruments	Primary use	Secondary use
Integrators		
Manager of managers		x
Management platforms		x
Element Management Systems for Wide Area Networks (WANs)		
Modems		x
Multiplexers		
Packet switches		
Fast packet switches		
ATM switches		
SMDS nodes		
ISDN nodes		
Mobile communication		
Matrix switches		
Operations support systems		
Element Management Systems for Local Area Networks (LANs)		
Bridges		x
Routers		
Brouters		
Repeaters		
Extenders		
Hubs		
Segments		
FDDI		
DQDB		
Monitors and Analyzers		
WAN monitor	x	
LAN monitor	x	
WAN analyzer	x	
LAN analyzer	x	
Network monitor	x	
Software monitor	x	
Security Management Systems		
Protection of systems and networks		
Protection of management systems		
Administration Instruments		
Documentation systems		
Modeling instruments	x	
Presentation tools		x
Report generators		x
Trouble-tracking tools		x
Software distribution tools		
Software licensing tools		
Database Tools		
Databases	x	
MIB browser		
Inquiry tools		x

Network management instruments	Primary use	Secondary use
Client Service Point Instruments		
Prediagnosis by phone		
Automated call distributor		
Voice mail		
E-mail		
Pager		
Console emulator		
Expert system		

theft of company telecommunications and computing equipment, as well as the software, data, or documentation associated with it. Major security management activities help accomplish the following goals:

- Minimizing the possibility of intrusion by using a layered defense system (e.g., a combination of policies and hardware and software solutions to build a uniform barrier to unauthorized users).

- Providing a means of quickly detecting unauthorized use and determining the original violation entry point, which should provide an audit trail of the violator's activity.

- Allowing the network manager to manually reconstruct any damaged files or applications and restore the system to the state just prior to the violation. This reconstruction feature helps minimize damage and allows system recovery.

- Ultimately, allowing the violators to be manually monitored and trapped by a network operations group, ending with a reprimand or prosecution.

More details about security management processes can be found in (Terplan, 1994). Typical instruments that support the process for security violation control are listed in Table 3.5.

TABLE 3.5 Typical Instruments for Security Management

Network management instruments	Primary use	Secondary use
Integrators		
Manager of managers		x
Management platforms		x
Element Management Systems for Wide Area Networks (WANs)		
Modems		x
Multiplexers		
Packet switches		
Fast packet switches		
ATM switches		
SMDS nodes		
ISDN nodes		
Mobile communication		

TABLE 3.5 Typical Instruments for Security Management (Continued)

Network management instruments	Primary use	Secondary use
Matrix switches		
Operations support systems		
Element Management Systems for Local Area Networks (LANs)		
Bridges		x
Routers		
Brouters		
Repeaters		
Extenders		
Hubs		
Segments		
FDDI		
DQDB		
Monitors and Analyzers		
WAN monitor		x
LAN monitor		x
WAN analyzer		
LAN analyzer		
Network monitor		
Software monitor		x
Security Management Systems		
Protection of systems and networks	x	
Protection of management systems	x	
Administration Instruments		
Documentation systems	x	
Modeling instruments		
Presentation tools	x	
Report generators		x
Trouble-tracking tools	x	
Software distribution tools		
Software licensing tools		
Database Tools		
Databases		
MIB browser		
Inquiry tools		
Client Service Point Instruments		
Prediagnosis by phone		
Automated call distributor		
Voice mail		
E-mail		
Pager		
Console emulator		
Expert system		

In general, the corporation's accounting policy defines the financial strategy and procedures that best meet overall corporate objectives. Collecting information, analyzing cost elements, and processing and verifying vendor bills are absolutely necessary. Establishing and defining chargeback policies and integrating network accounting management into corporate accounting, however, depends on corporate management. These functions are optional, but most corporations have some solutions in place.

Accurate accounting requires continuous and detailed monitoring, which often carries unnecessarily high overhead. Even bill verification is very difficult because of frequent tariff changes and differences between carriers, countries, and even states. Periodic monitoring techniques can be useful for establishing trends, range, and points in costing information. Continuous monitoring in this area is not feasible or cost-effective.

The present lifecycle of processing accounting-released information is too long; real-time or near real-time processing results are expected by most users. Accounting in LANs is virtually nonexistent today; most corporations consider LANs another element of the infrastructure. Only a few applications introduce the concepts of resource and usage accounting to interconnected LANs. More details can be found in (Terplan, 1994). Applications and tools that assist costing and billing are listed in Table 3.6.

TABLE 3.6 Typical Instruments for Costing and Billing

Network management instruments	Primary use	Secondary use
Integrators		
Manager of managers		
Management platforms		
Element Management Systems for Wide Area Networks (WANs)		
Modems		
Multiplexers		
Packet switches		
Fast packet switches		
ATM switches		
SMDS nodes		
ISDN nodes		
Mobile communication		
Matrix switches		
Operations support systems		
Element Management Systems for Local Area Networks (LANs)		
Bridges	x	
Routers	x	
Brouters	x	
Repeaters		
Extenders		
Hubs		
Segments		
FDDI		
DQDB		
Monitors and Analyzers		
WAN monitor	x	
LAN monitor	x	
WAN analyzer		

TABLE 3.6 Typical Instruments for Costing and Billing (Continued)

Network management instruments	Primary use	Secondary use
LAN analyzer		
Network monitor	x	
Software monitor	x	
Security Management Systems		
Protection of systems and networks		
Protection of management systems		
Administration Instruments		
Documentation systems	x	
Modeling instruments		
Presentation tools		x
Report generators		
Trouble-tracking tools		
Software distribution tools		
Software licensing tools	x	
Database Tools		
Databases	x	
MIB browser		
Inquiry tools		
Client Service Point Instruments		
Prediagnosis by phone		
Automated call distributor		
Voice mail		
E-mail		
Pager		
Console emulator		
Expert system		

Herman defines additional important processes, such as the following (Herman, 1994):

- capacity planning and resource sizing
- backup, restoration, and archive
- software distribution
- facilities planning
- service-level setting
- technology planning and upgrading
- contingency planning and disaster recovery
- billing and chargeback
- budgeting and capital planning
- security audits

3.4 Network Management Instruments

To support network management processes, functions, and human resources, a variety of instruments can be used. No single instrument addresses all functional areas, however, so it is in the best interest of the companies instrumenting network management to select the right mix of instruments. This segment of the book introduces the principal groups. Form 3.2 helps inventory existing and planned instruments. The principal groups are described in the following subsections.

FORM 3.2 Use of Network Management Instruments

Instruments	In use	Planned	Owner
Integrators			
Manager of managers			
Management platforms			
Element Management Systems for WANs			
Modems			
Multiplexers			
Packet switches			
Fast packet switches			
ATM switches			
SMDS nodes			
ISDN nodes			
Mobile communication			
Matrix switches			
Operations support systems			
Element Management Systems for LANs			
Bridges			
Routers			
Brouters			
Repeaters			
Extenders			
Hubs			
Segments			
FDDI			
DQDB			
Monitors and Analyzers			
WAN monitor			
LAN monitor			
WAN analyzer			
LAN analyzer			
Network monitor			
Software monitor			
Security Management Systems			
Protection of systems and networks			
Protection of management systems			

FORM 3.2 Use of Network Management Instruments (Continued)

Instruments	In use	Planned	Owner
Administration Instruments			
Documentation systems			
Modeling instruments			
Presentation tools			
Report generators			
Trouble tracking tools			
Software distribution tools			
Software licensing tools			
Database Tools			
Databases			
MIB browser			
Inquiry tools			
Client Contact Point Instruments			
Prediagnosis by phone			
Automated call distributor			
Voice mail			
E-mail			
Pager			
Console emulator			
Expert system			

Management integrators. These are instruments to supervise and control multiple element management systems and managed objects using a variety of management applications, standardized user interfaces, and multiple gateways to communicate with simple network management protocol (SNMP), common management information protocol (CMIP), and proprietary agents.

WAN element management systems. These are instruments to supervise and control a homogeneous family of managed objects in the WAN area, such as modems, multiplexers, packet switches, fast packet switches, asynchronous transfer mode (ATM) switches, matrix switches, switched megabits data service (SMDS) nodes, integrated services digital network (ISDN) nodes, mobile communication units, and operations support systems, using a variety of device-specific management applications, standardized user interfaces, and multiple gateways to communicate with SNMP, CMIP, and proprietary agents.

LAN element management systems. These are instruments to supervise and control a homogeneous family of managed objects in the LAN area, such as Ethernet segments, token ring segments, fiber-distributed data interface (FDDI) segments, distributed queue dual bus (DQDB) segments, bridges, routers, brouters, extenders, repeaters, and hubs using a variety of device-specific management applications, standardized user interfaces, and multiple gateways to communicate with SNMP, CMIP, and proprietary agents.

WAN monitors. These are special-purpose instruments to continuously measure key fault and performance indicators in various locations of the WANs.

LAN monitors. These are special-purpose instruments to continuously measure key fault and performance indicators in various locations and segments of the LANs.

WAN analyzers. These are special-purpose instruments to diagnose and troubleshoot faults and analyze bottlenecks in various locations of the WANs.

LAN analyzers. These are special-purpose instruments to diagnose and troubleshoot faults and analyze bottlenecks in various locations and segments of the LANs.

Network monitors. These are hardware-based information extraction, processing, and reporting instruments that use published data communication interfaces to target the continuous supervision of complete networks. Network monitors use sampling technology.

Software monitors. Software monitors are software-based information extraction, processing, and reporting instruments, using private and published software interfaces and user exits, targeting the continuous supervision of large networking segments using eventing or sampling technology.

Security management systems. These instruments help with authorization and authentication of the usage of systems and networking resources to protect the networks and systems management products in general, as well as protect against viruses.

Administration instruments. Administration instrument is a generic term for all the instruments responsible for administering components of networks and systems. Included are tasks such as asset management, software distribution, documentation, and information distribution services.

Documentation tools. Documentation tools are instruments tightly coupled with administration and configuration management tools. The principal goal of these tools is to maintain the actual status of managed objects, their connectivity, and their dynamic indicators.

Presentation tools. Presentation tool is a generic term for all instruments that process data stored in various databases so that data can be presented to users in various forms, such as lists, tables, graphs, pie charts, diagrams, and other user-customizable forms.

Trouble-tracking tools. Trouble-tracking tools are special-purpose applications to open, dispatch, track, and close trouble tickets. These tools include processing, notification, and reporting features.

Modeling instruments. These are instruments to predict the future service quality and use of networking and systems resources by using simulation, emulation, and applied-queuing technology.

Databases. This generic group of instruments supports the maintenance of configuration, performance, troubles, and asset-related data in relational or object-oriented schemas. Products of this category include the database management system, but populating the tools remains the responsibility of the users.

Management information base tools. This group of instruments helps populate management information bases (MIBs) and retrieve information from MIBs using a browser.

Console emulator. The console emulator is a special-purpose instrument that represents multiple element management systems, monitors, and analyzers using Windows on state-of-the-art workstations, allowing more centralized and consolidated management. If necessary, Windows offers the opportunity to operators to directly access the system experiencing the problems.

Expert systems. Expert systems are instruments that automatically determine and diagnose complex management-related problems using rules and input from various measurement and management tools. Both offline and online operational alternatives are supported. Fault-tolerant and self-healing systems are usually grouped into artificial intelligence. Expert systems make decisions based on the detected status of facilities and equipment, but are not typically able to learn.

3.5 Network Management Protocols

Network management solutions are greatly affected by emerging standards. Standards can include proprietary solutions, such as network management vector transport (NMVT) or message service unit (MSU) from IBM, which are de facto standards for transmission control protocol/internet protocol

(TCP/IP)-based networks, including SNMP versions 1 and 2 (SNMPv1 and SNMPv2), and open solutions based on open system interconnection (OSI) and International Telephone and Telegraph Consultative Committee (CCITT) recommendations, including the CMIP and telecommunication management network (TMN).

In addition, organizations promote application programming interfaces (APIs) between standards and frameworks that include these standards. These groups include OMNIPoint from the Network Management Forum, distributed management environment (DME) from the Open Systems Foundation, ATLAS from Unix International, and government network management profile (GNMP) from the U.S. Government OSI Profile (GOSIP). Protocol standards such as SNMP and CMIP are more mature than API standards such as DME and desktop management interface (DMI) from the desktop management task force (DMTF).

Finally, groups with special interests on particular technologies have also been established to support ATM, FDDI, and desktop management.

The use of these standards helps accelerate widespread acceptance of management applications and provide some measure of future proofing. Specific management functional areas and services have been defined by the International Standards Organization (ISO) in documents concerning OSI. The primary OSI management protocol is CMIP. Performance of these standards, however, has not yet been widely proven in the marketplace. OSI management encompasses both WAN and LAN management, but the estimated overhead scares both vendors and users away.

The Internet Engineering Task Force (IETF) in the United States has also defined standards for the management dialog in networks using the TCP/IP protocol. Chief among these standards is SNMP. The market rapidly embraced SNMP after its introduction in 1989, and the protocol is now a common denominator for management and will continue to be so for the remainder of this decade.

Both SNMP and CMIP are defined as *application-layer protocols* that use the underlying transport services of the protocol stacks. The management data exchanged between managers and agents is called the *management information base*, or MIB. Data definitions must be understood by both managers and agents.

The manager-agent paradigm is central to both SNMP and CMIP architectures. A *manager* is a software program housed within the management station. The manager has the ability to query agents using various SNMP commands. The management station also interprets MIB data, constructs views of the systems and networks, compresses data, and maintains data in relational or object-oriented databases.

The MIB is a virtual database of managed objects, accessible to an agent and manipulated via SNMP to achieve network management. Recently, remote monitoring information has been incorporated into a standard MIB, known as RMON, which is an abbreviation for Remote MONitoring. Figure 3.7 shows a generic SNMP-based structure.

Figure 3.8 shows the overview of processes and communication links between managers, agents, subagents, and managed objects. These processes are almost the same for all types of protocols; differences occur in respect to the initiator of the communication exchange, contrasting polling- and eventing-based technologies. For standardizing the manager-agent dialogue, several issues must be carefully considered.

How will the management information be formatted and how will the information exchange be regulated? This question is actually the protocol definition problem.

How will management information be transported between manager and agent? The OSI standards employ the OSI protocol stack, and the TCP/IP standards use the TCP/IP stack, but this solution is no longer the only one. Several protocols demonstrate the emerging independence of management protocols from underlying protocol layers, including CMIP over TCP/IP (CMOT), CMIP over logical link control (CMOL), and Common Management Information protocol Over SNA (CMOS).

SNMP and OSI employ the concept of *managed objects*. Management objects are defined by their attributes, or operations that can be performed

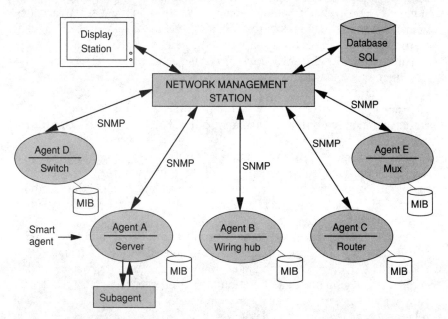

MIB   = management information base
SNMP = simple network management protocol

Figure 3.7 Generic SNMP structure

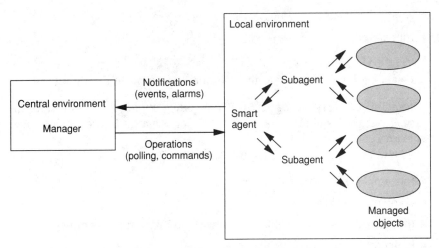

Figure 3.8 Communication links among manager, agents, subagents, and managed objects

on them, and the notification that can result. The set of managed objects in a system, together with its attributes, constitute that system's MIB. In addition to MIBs, the structure of management information (SMI) defines the logical structure of OSI or SNMP management information. The MIB can be extended to include private variables for describing devices or components offered by vendors or developed by users.

In terms of SNMP, the following trends are expected: SNMP agent-level support, already widely used, will be provided by an even greater number of vendors. SNMP manager-level support will be provided by only a few leading vendors in the form of several widely accepted platforms. Management platforms provide basic services, leaving customization and the development of additional applications to vendors and users. The RMON MIB will help bridge the gap between the limited services provided by management platforms and the rich sets of data and statistics provided by traffic monitors and analyzers.

RMON defines the next generation of network monitoring with more comprehensive network fault diagnosis, planning, and performance tuning features than any current monitoring solution. It uses SNMP and its standard MIB design to provide multivendor interoperability between monitoring products and management stations, allowing users to mix and match network monitors and management stations from different vendors.

The strengths of SNMP include the following: (Terplan, 1994)

- agents widely implemented
- simple to implement
- minimal agent-level overhead

- good polling approach for LAN-based managed objects
- robust and extensible
- best direct manager-agent interface
- meets a critical need—available and implementable at the right time

In addition, version 2 of SNMP is now being implemented in vendor offerings, offering significant security and performance improvements over version 1.

The weaknesses of SNMP include the following (Terplan, 1994):

- too simple, does not scale well
- no object-oriented data view
- unique semantics make integration with other approaches difficult
- high communication overhead due to polling
- many implementation-specific (private MIB) extensions
- no standard control definitions
- small agent (one agent per device) might be inappropriate for systems management

In contrast, the strengths of CMIP include the following (Terplan, 1994):

- general and extensible object-oriented approach
- support from the telecommunications industry and international vendors
- support of manager-to-manager communications
- support of a framework for automation

The weaknesses of CMIP include the following (Terplan, 1994):

- complex and multilayered
- high overhead
- few CMIP-based management systems shipped
- few CMIP-based agents in use

A consortium of vendors and users is seeking to overcome market resistance to CMIP by promoting a comprehensive integration plan, called OMNIPoint. These open management interoperability (OMNI) points offer an industry solution to a worldwide problem of managing heterogeneous networks. An OMNIPoint is an agreed-upon approach to the development and procurement of interoperable management systems and their components. OMNIPoints include standards and profiles, industry agreements, object definitions, and supporting technologies (Network Management Form 1993).

The goals of OMNIPoint are to establish intercepts at specific intervals of standard development; to standardize migration rules; to reduce upgrade costs, and to foster technological changes as early as reasonable. OMNIPoint deliverables concentrate on guidelines for purchase and development decisions, library services such as object definitions and object catalogues, specification for standards, and test procedures and tools.

OMNIPoint clarifies communications between management systems using CMIS/X.700 and managed objects using SNMP, CMIS/X.700, and proprietary protocols. The transport itself is supported by OSI, TCP/IP, user datagram protocol internet protocol (UDP/IP), LAN logical link control (LCC), and proprietary protocol stacks. The internal interfaces include APIs for the user interface, the MIB, the communication with CMIP and SNMP agents, and network management applications. OMNIPoint does not differentiate between WAN and LAN management. It offers the following services:

- security management

- problem management

- status management

- alarm and event management

The Network Management Forum is also coordinating its OMNIPoint program with the telecommunications management network (TMN) architecture. TMN can be defined as an organized architecture for achieving interconnection between various types of operations support systems and telecommunications equipment. TMN incorporates a functional, information, and physical architecture. The *q-interface* plays a key role with TMN; it is an interface applying the *q-reference point*, where the q-reference point is the point between two physically separate functional blocks. The q-interface is defined by the protocol it uses and the information carried across it.

The TMN management services are different from those recommended by OSI bodies; in the case of TMN, products have been developed for switching management, tariff and charging administration, management of customer access, materials management, management of transport networks, management of circuits and routes, and staff work scheduling.

TMN plays an important role in offering the standard umbrella for all networks operated by phone companies and network services providers. Conformance to TMN must be evaluated if benchmarking telecommunication providers.

As more network management application software is produced, it becomes important to define standardized APIs so that the software can be easily ported to different management platforms and that software developed by different vendors can be easily combined on a single platform. The management platforms provide a standardized environment for developing and implementing applications, and they also separate management-application software from the usual system-level services.

Many independent companies now offer network management applications designed to provide real multivendor solutions while taking advantage of the system-level services of platforms. This approach allows third-party vendors to concentrate on their specific hardware and software. Users can focus on the customization and fine tuning of management applications. Figure 3.9 shows how applications, platforms, and protocol interface modules work together.

Network management benchmarks should also include an analysis of what types of network management applications are in use or under consideration. Management application groups include the following (Terplan, 1994):

- platform extensions
- device-specific applications
- process-specific applications
- service-specific applications
- systems management applications
- traffic monitoring and analysis applications
- application development toolkits

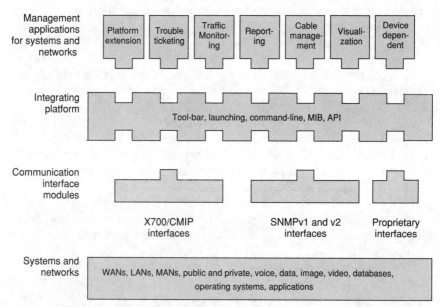

Figure 3.9 Platforms, applications, and protocol interfaces

It is also important to analyze the depth of integration between the applications and network management products. The most popular alternatives today include the following (Terplan, 1994):

- application launching
- command-line interface
- event-level integration
- database or MIB-level integration
- standardized application programming interfaces

In network management applications and their integration, industry analysts expect revolutionary changes in the near future. Standards work together in various combinations. SNMP and CMOL are very important for manager-agent-subagent communications. CMIP and SNMPv2 can serve to support peer-to-peer manager communications. For user interfaces, vendors are implementing Motif, X-Windows, Openlook, and DOS/Windows. In distributed implementations, remote procedure calls (RPC) or object invocation techniques are used. OMNIPoint implements X/Open management protocol (XMP) to hide other protocols, such as CMIP and SNMP, from users.

Form 3.3 supports the benchmarking team with the inventory of established and planned network management protocols. Specific network management protocols are also surveyed.

FORM 3.3 Network Management Protocols

Network Management Protocols	Established	Planned	Comments
Proprietary protocols			
IBM			
NMVT			
MSU			
DEC			
Novell			
Others			
Standard protocols			
CMIP			
SNMPv1			
SNMPv2			
CMOL			
CMOT			
XMP			
RMON			
OMNIPoint			
Edge			
SQL (Structured Query Language)			
RPC (Remote Procedure Call)			

3.6 Human Resources Supporting Network Management

Network management is traditionally part of the management information systems (MIS) organization. This fact has many contradictory results. For example, MIS is still very much data-oriented because of the traditionally hierarchical processing solutions. MIS does not see the necessity of distributing processing and management functions, nor does it deal with voice and video communications. MIS tends to feel responsible for the logical network management with roots in network operating systems and telecommunication access methods. MIS thus treats infrastructure-related questions of cabling, environmental control, wiring hubs, uninterrupted power supply, and similar items as routine and therefore of a low priority.

On the other hand, telecommunications departments offer some benefits, including the installation services for phone sets, cabling and its maintenance, private branch exchange (PBXs) maintenance , and liaisons with network services providers.

In addition to the conflicting capabilities of MIS versus telecommunications, business units that represent users want LANs designed, planned, and operated with mission-critical client/server-applications running on them. By careful combination of motivational techniques, these three different directions of managing networks can be unified.

One method is to organize the network management team in accordance with network management functions as outlined earlier in this chapter. If this structure is too detailed, a higher level of functional groups must be introduced. Usually, companies group network management functions around operations, administration, and planning and design. Table 3.7 shows this high-level grouping of functions.

The allocation is very straightforward following the traditional method of organizing functions. Figure 3.10 shows typical jobs of subject-matter experts and persons allocated to each high-level functional area, such as operations, administration, and design and planning.

For each activity area, typical profiles must be developed and implemented. Such job descriptions or profiles include job responsibilities, job contacts, expected skill levels, expected educational backgrounds, and salary ranges. Chapter 4 offers examples for all job titles identified in Form 3.4.

The process of building and expanding the network management team is complicated by a number of factors, including the following most important ones:

- scarce technical resources
- no standard job descriptions and responsibilities
- few specific academic training programs
- rapidly changing networking technology
- changes in network management instrumentation

- shorter career paths
- few upwardly mobile alternatives

These factors complicate the hiring process by making it very difficult to write job descriptions and analyze candidates' background materials. The following is a list of recommended criteria when hiring network management staff.

Identify team members. The earlier segments of this chapter gave an overview of principal network management functions. Depending on the size of

TABLE 3.7 Allocation Matrix of Functional Groups to Organizational Units

| | Organizational units | | |
Functional groups	Operations	Administration	Design and planning
Client contact point	x		
Operations support	x		
Fault tracking	x		
Change control		x	
Planning and design			x
Finance and billing		x	
Implementation and maintenance	x		(x)
Fault monitoring	x		
Performance monitoring	x		
Security management		x	
Systems administration		x	

x   Primary allocation
(x)   Secondary allocation

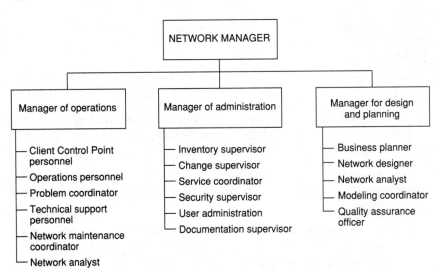

Figure 3.10   Organizational structure of network management

FORM 3.4 Human Resources that Support Network Management

Job titles	Established	Planned	Other job titles
Network manager			
Network administration supervisor			
Inventory and assets coordinator			
Cable management administrator			
Change coordinator			
Problem coordinator			
Order processing and provisioning coordinator			
Network operations control supervisor			
Client contact point operator			
Network operator			
Network technician			
Network performance supervisor			
Network performance analyst			
Database analyst			
Database administrator			
Modeling coordinator			
Security management supervisor			
Security officer			
Security auditor			
Security analyst			
Finance and control supervisor			
Costing specialist			
Accounting clerk			
Charging specialist			
Documentation supervisor			
Network maintenance supervisor			
Quality assurance officer			
Inspector			
Network implementation coordinator			
Service-level coordinator			
Supervisor of standards			
User administrator			
Design and planning supervisor			
Business planner			
Technology analyst			

the network, the human resources demand can be computed. After subtracting the available staff from the total demand for each functional area, the demand on new hires can be quantified.

Recruit candidates. Advertisements, conferences, headhunters, and individual contacts to colleges, universities, and other companies help find candidates.

Establish interview criteria. Guidelines and evaluation criteria must be set prior to starting the interviews. To keep investment for both parties low, written applications must be filtered carefully. Occasional phone conversations can fill existing gaps. Invitations to personal interviews should be sent out to candidates whose applications match company expectations.

Hire properly qualified candidates. Hiring must be for mutual benefit, rather than just to fill the job. Future turnover can be avoided if this step is followed.

Assign or reassign responsibilities. Static job descriptions should serve as a guideline only. Within this framework, more dynamic descriptions with rotations are necessary.

Institute performance evaluations. Periodic reviews are most widely used. If possible, upward performance appraisals should be agreed upon as early as possible in the team-building phase.

Promote openness and handle complaints. To emphasize team spirit, opinions and even complaints must be encouraged on behalf of network management supervisors. The employees must have the feeling that their comments and suggestions are handled at the earliest convenience of managers.

Resolve personnel problems quickly. To avoid tensions within the network management organization, problems must be resolved to mutual benefit as quickly as possible. The reward system must provide opportunities to do so. Often, a visibility of how the reward system works can resolve problems automatically.

Institute systematic training and development programs. Systematic education should include training for network management functions, network management instruments, and personal skills. A curriculum in coordination with vendors and educational institutes guarantees high quality and employee satisfaction.

Regularly interface network management staff with users. To promote a mutual understanding of working conditions and problems, both parties should exchange views and opinions. The level of formality can vary from very informal to very formal; in the second case, written service-level agreements are evaluated.

Evaluate new technologies. As part of the motivation process, network management renovation opportunities must be evaluated continuously. This process includes new management platforms, new technologies of distribution, feasibility of new and existing solutions, new monitors, changes in de facto and open standards, simplification of management processes, changes in the offerings of leading manufacturers, and monitoring the needs of users. Thus, enrichment of lower-level jobs can easily be accomplished.

To keep the network management team together, expectations of employers and employees must match to a certain degree. The individual/organization contract is termed *psychologically* because much of it is often unwritten and unspoken. Several reasons exist for keeping it unwritten in the beginning:

- Both parties might not be entirely clear about their expectations and how they wish them to be met. They might not want to define the contract until they have a better feel for what they want.

- Neither of the parties is aware of expectations. For example, organizations are hardly able to define the term *loyalty*.

- Some expectations might be perceived as so natural and basic that they are taken for granted, such as not stealing and putting in an honest day's work for a day's pay.

- Cultural norms might inhibit verbalization, which is very important with multinational companies that hire employees from various countries.

At any given time, there are fulfilled and unfulfilled expectations; however, each party must have a minimum acceptance level of fulfillment. If either party concludes that the fulfillment of its needs is below this minimum level, it views the contract as having been violated.

Turnover in network management can be very disadvantageous for maintaining service levels to end users. Corporate and business unit management should try to avoid greater than average turnover by implementing rewards to satisfy employees. Gaining satisfaction with the rewards given is not a simple matter; it is a function of several factors that organizations must learn to manage.

The individual's satisfaction with rewards is, in part, related to what is expected and how much is received. Feelings of satisfaction or dissatisfaction arise when individuals compare their input (knowledge, skills, experience) to the output (mix of rewards) they receive.

Employee satisfaction is also affected by comparisons with other people in similar jobs and organizations. People weigh various inputs and outputs differently in their comparisons. They tend to weigh their strong points more heavily, such as certain skills or a recent performance peak. Individuals also tend to correlate their own performance compared to the rating they receive from their supervisors. The problem of unrealistic self-ratings exists partly because supervisors in most organizations do not offer a candid evaluation of their subordinates' performance.

Employees often misperceive the rewards of others; their misperception can cause employees to become dissatisfied. Evidence shows that individuals tend to overestimate the pay of colleagues doing similar jobs and underestimate their colleagues' performance.

Overall satisfaction results from a mix of rewards rather than from any single reward. Rewards fall into two principal categories: extrinsic and intrinsic. Extrinsic rewards come from the employer in the form of compensation, benefits, job security, training, promotions, effective network management instruments, and recognition. Intrinsic rewards come from performing the task itself and can include job satisfaction, a sense of influence, quality of environment, and quality of assignment. The priority of extrinsic and intrinsic rewards depends on the individual person. The following list gives a frequently seen priority sequence to keep the network management team together:

1. *Compensation:* Payment is still the most important motivation factor. Organizations try to use a number of person-based or skill-based compensation techniques, combined with the dependence of sales revenues of the larger organization, if applicable. Pay is a matter of perception and values and can often generate conflict.

2. *Benefits:* Benefits take special forms, depending on the employer's business; such as a company car, life insurance, lower interest rates, or housing. The cost of benefits at companies can be as high as 35 to 45 percent per dollar of pay.

3. *Job security:* Seniority with job assignments is a very valuable management practice, particularly when the economy is stressed. Job security policies include retirement plans, options for early retirement, and agreements of nonlayoff. Job security packages are more advanced in Europe and Japan than in the United States.

4. *Recognition:* Recognition can come from the organization or from fellow employees. The periodic form of recognition is the performance appraisal conducted by the supervisor. A relatively new form, the so-called *upward appraisal* is considered a form of subordinates' recognition. It is difficult to implement because most managers do not want to be evaluated by their subordinates. For the subordinates, however, it is the forum for communicating their ideas for improvement.

5. *Career path and creation of dual ladders:* To keep motivation high, managerial and technical assignments must be compensated equally. Promoting technically interested persons into managerial positions might not have desired results; these persons are usually high in affiliation motivation and low in power motivation. Helpful activities include career counseling and exploration, increased company career opportunity information, improved career feedback, enhanced linear career moves, slower early-career advancement, and enrichment of lower-level jobs with more challenges.

6. *Effective training:* This type of motivation helps keep the specific and generic knowledge of the employees at the most advanced level. Three to six weeks of training and education annually is considered adequate in a dynamically changing network management environment.

7. *Quality of assignments:* Job descriptions are expected to give the framework for expectations. Dynamic job descriptions can help avoid monotony and promote job rotation. The client contact point, systems administration, and change control positions can be rotated periodically.

8. *Use of adequate tools:* Better instrumented networking environments facilitate the jobs of the network management staff, increase the service quality to users, and improve the image of the network management organization. At the same time, persons working with advanced tools are proud of their special knowledge and their employer. They are highly motivated to continue with the company.

9. *Realistic performance goals:* As part of dynamic job descriptions and job rotation, realistic performance expectations can help stabilize the position of the network management team. Management must find the balance between quantifiable and nonquantifiable goals. Average time spent on trouble calls, response time to problems, time of repair, and end-user satisfaction or dissatisfaction are examples for both types of goals.

10. *Quality of environment:* This parameter is more or less a generic term expressing the mix of network management related instruments, pleasant working atmosphere, comfortable furniture, adequate legroom, easy access to filing cabinets or hypermedia, acceptance of opinions on shortcomings, and team spirit.

11. *Employee control:* Despite high team spirit, individuals need certain levels of control that can only be determined by managerial skills. Depending on the person, positive or negative motivation or a combination of both can work best.

The preceding list concentrates on key motivation alternatives only. Many more options exist. To find the optimal combination for individual installations, a human resources management audit is recommended.

Forms 3.4 and 3.5 address these items by offering a number of job titles within the network management organization and by asking special questions about the skill levels of the network management staff.

3.7 Summary

Critical success factors help streamline all network management related activities on the most important issues impacting the overall performance of network management. Processes, instruments, protocols, and human resources have been outlined in some depth. Five different forms help define targets and present status for network management of the company under benchmarking. These forms can be used as the primary documentation for questionnaires, interviews, observations, and measurements. In addition, more indicators are defined and quantified in the following chapters.

FORM 3.5 Responsibility/Skill Matrix for Network Management Areas

Client Contact Point

Functions	Functional knowledge	General telecom	In-depth telecom	Instrument knowledge	Personal communication	Project management
Receiving problem reports						
Handling calls						
Handling enquiries						
Receiving change requests						
Handling orders						
Making service requests						
Opening and referring trouble tickets						
Closing trouble tickets						

Operations Support

Functions	Functional knowledge	General telecom	In-depth telecom	Instrument knowledge	Personal communication	Project management
Determining problems (using trouble tickets)						
Diagnose problems						
Take corrective actions						
Repairing and replacing						
Referencing to third parties						
Backing up and reconfiguration						
Recovering						
Logging events and performing corrective actions						

FORM 3.5 Responsibility/Skill Matrix for Network Management Areas (Continued)

Fault Tracking

Functions	Functional knowledge	General telecom	In-depth telecom	Instrument knowledge	Personal communication	Project management
Tracking manually reported or monitored faults						
Tracking progress and escalation of problems if necessary						
Distributing information						
Referring problems						

Change Control

Functions	Functional knowledge	General telecom	In-depth telecom	Business administration	Personal communication	Project management
Managing, processing, and tracking of service orders						
Routing service orders						
Supervising the handling of changes						

Planning and Design

Functions	Functional knowledge	General telecom	In-depth telecom	Business administration	Personal communication	Project management
Analyzing needs						
Projecting application load						
Sizing resources						
Authorizing and tracking changes						
Raising purchase orders						
Producing implementation plans						
Establishing company standards						
Maintaining quality assurance						

Finance and Billing

Functions	Functional knowledge	General telecom	In-depth telecom	Business administration	In-depth tariff information	Project management
Asset management						
Costing services						
Client billing						
Usage and outage collection						
Calculation of rebates to clients						
Bill verification						
Software license control						

FORM 3.5 Responsibility/Skill Matrix for Network Management Areas (Continued)

Implementation and Maintenance

Functions	Functional knowledge	General telecom	In-depth telecom	Practical experiences	Personal communication	Project management
Implementing change requests and work orders						
Maintaining resources						
Inspection						
Maintaining the configuration database						
Provisioning						

Fault Monitoring

Functions	Functional knowledge	General telecom	In-depth telecom	Instrument knowledge	Personal communication	Project management
Monitoring system and network for proactive problem detection						
Opening additional trouble tickets						
Referring trouble tickets						

Performance Monitoring

Functions	Functional knowledge	General telecom	In-depth telecom	In-depth instrument knowledge	Creativity and patience	Project management
Monitoring system and network performance						
Monitoring service-level agreements						
Monitoring third-party and vendor performance						
Optimizing, modeling, and tuning						
Reporting on usage statistics and trends						

Security Management

Functions	Functional knowledge	General telecom	In-depth telecom	Security issues	Creativity	Patience
Threat analysis						
Administration						
Detection						
Recovery						
Protecting the management systems						

FORM 3.5 Responsibility/Skill Matrix for Network Management Areas (Continued)

| | Systems Administration | | | | | |
Functions	Functional knowledge	General telecom	In-depth telecom	Instrument knowledge	Personal communication	Project management
Software version control						
Software distribution						
System management						
Administration of user-definable tables						
Local and remote configuration of resources						
Management of names and addresses						
Applications management						

4

Benchmarking Indicators

4.1 Introduction

To evaluate the status of network management functions, processes, instruments, and human resources, sufficient data needs to be collected. Data collection can take several forms. This chapter includes questionnaires that address various items of networking and network management. The preliminary questionnaire asks for information about the company background, applications, transmission facilities, networking equipment, personnel, costs, and basic network management. These questions are sent out prior to starting the benchmarking. Depending on the company size, two to four weeks of lead time is needed to complete the form.

The second questionnaire forms the basis for the on-site interviews. The focus is on company investments and the organizational structure; network management functions and problems; the implementation of processes, protocols, and instruments; and network management directions. The answers are not always easy to obtain; an interactive working style is recommended, along with a targeted completion time of 3 to 10 working days. It is desirable to receive as much detailed documentation as possible.

The third group of questions offers great detail in analyzing single network management processes and functions organized by business area. The questions asked about each function are very similar to each other; however, the depth of the answers is very different. This questionnaire is very time consuming to complete. Usually, not all functions are investigated in depth, just those supported by the company. In most cases, not all functions are supported. The schedule for this activity follows after Form 3.1 is completed.

Most companies possess a lot of information but, unfortunately, that information sometimes is in "data jails;" it is hidden in files, such as trouble tickets, asset inventories, purchase orders, change requests, etc. Data collection activities include searching for this hidden data. This chapter also includes indicators including generic, organizational, network management process based, and cost. These indicators can be filled out directly by users or indirectly by the benchmarking team using other data sources. In both cases, the use of spreadsheets for consolidating data is very useful.

Other information sources include logs that represent the accurate observation of operations. Logs can be prepared for all principal network management processes under consideration. They give the sequence of certain activities, using time windows of 30 minutes. Logs can be organized by shifts; more detailed logs are recommended for the shift takeover. In addition to logs, special reports can be used that are part of documented procedures of network management control centers.

Highly accurate data can be collected using monitors in various networking segments. Depending on the server and network operating systems of targeted companies, built-in software monitors can collect the requested information. In other cases, hardware or software probes can be installed in networking segments to collect information for a certain period of time. The final decision on monitoring depends on the results of Form 3.2, where instruments in use and in planning were surveyed.

4.2 Questionnaires

This segment describes the necessary questions organized for off-premise and on-premise interviews. These questions should be mailed in advance to the company and its distributed locations to give it time to prepare the answers. The benchmarking team must not be surprised and frustrated when the answers have not been prepared. In most cases, interviews fill the gaps left from the questionnaires. Because of the length of the individual questionnaires, they have been provided in full as appendices.

Preliminary questionnaire

The preliminary questionnaire, found in Appendix A, asks about the company background, applications, transmission facilities, networking equipment, personnel, costs, and basic network management. These questions are sent out prior to the start of the benchmarking. Depending on the company size, two to four weeks are provided for the company to complete this questionnaire. This questionnaire must be completed prior to the on-site benchmarking interviews.

On-site interview questionnaire

The second questionnaire, shown in full in Appendix B, is the basis of the on-site interviews. The focus is on investments and organizational structure; network management functions and problems; the implementation of processes, protocols, and instruments; and the direction of network management. It is not always easy obtaining the answers; an interactive working style is recommended. A targeted completion duration is 3 to 10 working days. Obtain as much detailed documentation as possible during the interviews.

In-depth questionnaire for network management functions

The third group of questions, provided in Appendix C, provides great detail for analyzing single network management processes and functions organized by business area. The questions asked about each function are very similar. The depth of the answers, however, is very different. This questionnaire is very time-consuming. Usually, not all functions are investigated in-depth, just those supported by the company. In most cases, all functions are not supported. The schedule for this activity follows after Form 3.1 is completed.

Before beginning this task, a number of definitions are provided to help better understand the questions. Answering these questions needs a lot of time; the benchmarking team needs to work together with the company team to obtain thorough answers.

Functional groups of network management: summary functions of a business or organizational unit

Network management function: a clearly defined group of management activities

Network management activity: the smallest unit of a task within a network management function

Network management application: it addresses single activities, single functions, or a group of functions.

4.3 Observations and Activity Logs

Observations and activity logs are very important information sources. In both cases, manual activities dominate. Human errors cannot be excluded, but observations are sometimes the only information source for the benchmarking team.

Observations

Not only does existing documentation, the monitoring of results, and answers to interview questions contribute to successful benchmarks, but so do actual observations of operations. Observations complement the other activities of data collection. If appropriate documentation is available, observations can be reduced to a reasonable minimum of taking notes about how processes and procedures are implemented.

Three areas—client contact point, operations, and shift takeover—need to be concentrated on. The type of observations need for each area is described in the following subsections.

Client contact point. Improvisations are expected when observing the client contact point because the team must interface with various users who have very different technical and technological backgrounds. Observations include the following:

- how quickly phones are answered
- how polite the phone conversation is
- how the staff uses other tools, such as screens, files, faxes, and E-mails
- whether the right sequence of questions about symptoms is asked during the conversation
- how efficiently help features on the screen are used
- whether staff can help troubleshoot simple problems

It is also important to observe the daily and weekly workload fluctuation at the client contact point. The following observations should occur:

- how many calls are handled during peak hours
- activities at the client contact point during lightly loaded hours, concentrating on
  ~report generation
  ~trouble tracking
  ~information distribution to users
- what instruments are used
- the atmosphere, including noise levels, fresh air, efficiency of air conditioning and heating, and conversation between team members

It is also necessary to observe whether external interruptions are frequent or not. If interruptions are frequent, identify which group interfaces the most with the client contact point and the subject of the information exchange.

Operations. Operations is another area where human interaction is very likely. Interactions are expected between management and users, management and vendors, and the management team members.

Observations should emphasize the following items:

- what tools are used during problem determination and diagnosis
- how efficiently these tools are used by the management team
- if problems are jointly diagnosed at one site or if multiple sites are involved
- whether problem diagnosis software is exchanged between locations
- how frequently and how intensively operations manuals are used
- how many management systems one person operates
- when vendors and outside help are contacted
- how efficient pager systems are
- how quickly vendors and contacted outside help arrive to troubleshoot the problem
- whether remote maintenance is used and if so, whether special security protection is implemented
- whether information exchange is supported between telecommunications services providers

Shift takeover. Network management centers usually operate 7 days per week, 24 hours per day. It is extremely important to guarantee a smooth operation, and thus the takeover of shifts is very important. Observations should concentrate on the following:

- how much time the takeover takes
- whether the new team arrives on time and the old team leaves on time
- similarities and differences between takeovers
- the documentation given to the arriving team
- the verbal explanations of the open problems
- team spirit during turnover
- whether the trouble-ticketing system is used as part of the documentation
- the quantities of open trouble tickets at the end of the shift in relation to resolved trouble tickets
- whether graphical support is used to more clearly explain problems
- how many persons are involved in the shift-takeover procedure
- where the takeover actually takes place, such as the office of the supervisor, the main operating area, or the cafeteria
- how quickly the arriving team deals with the still-open problems

Observations are no replacement for accurate information provided in writing or during interviews with the network management staff. Observations do, however, complement other information sources. Feedback about the real team spirit is provided with observations, along with the atmosphere in the network management center.

TABLE 4.1   Activity Log

Time	User	Process	Function/Activity
First shift			
06:00–07:00			
07:00–08:00			
08:00–09:00			
09:00–10:00			
10:00–11:00			
11:00–12:00			
12:00–13:00			
13:00–14:00			
Second shift			
14:00–15:00			
15:00–16:00			
16:00–17:00			
17:00–18:00			
18:00–19:00			
19:00–20:00			
20:00–21:00			
21:00–22:00			
Third shift			
22:00–23:00			
23:00–24:00			
24:00–01:00			
01:00–02:00			
02:00–03:00			
03:00–04:00			
04:00–05:00			
05:00–06:00			

Activity logs

Activity logs record all details of operating the network. Usually, logs include the time window, the user, and the process, as well as the functions in some detail. Table 4.1 shows an example of the format of an activity log.

4.4 Indicators for Network Management

To quantify and compare the performance of various operations, quantifiable indicators are very important. Depending on the quality of raw data, indicators can be directly populated or indirectly computed using uncorrelated data. This section uses many samples; some populated with practical data originated in several companies and some uncorrelated.

The following indicator groups have been defined:

- generic indicators
- organizational indicators
- specific network management process indicators
- cost indicators

Each are described in the following subsections.

Generic indicators

The generic indicators are

- quality of documentation—the readability, depth, actuality, completeness, and look
- user satisfaction—user satisfaction cannot always be quantified, but it can consolidate many subindicators of performance and the responsiveness of the network management organization
- attended operations—this indicator quantifies days and hours when network management service is available to clients
- scalability of network management functions—this indicator reviews the scale-up capabilities of the network management organization, its functions, processes, tools, and coverage of services
- service quality of network management services—this indicator shows how the network management organization performs
- changability of service quality indicators—this indicator expresses the capability of clients and network management staff for changing service quality indicators if necessary

Forms 4.1 through 4.6 are shown for each of the generic indicators.

FORM 4.1 Quality of Documentation

Quality	Readability	Depth	Actuality	Completeness	Presentation
Physical network structure					
Logical network structure					
Backup plans					
Escalation procedures					
Disaster recovery plans					
Fault recovery plans					
Job descriptions					
Organization charts					
Maintenance contracts					
Contracts with external companies					
Operational manuals					
Security guidelines					
Service-level agreements					

FORM 4.2 User Satisfaction

| Time window | Level of satisfaction | | | |
	Excellent	Good	Fair	Poor
First shift				
Second shift				
Third shift				
Holidays				
Weekends				

FORM 4.3 Attended Operations

| Indicators | Level of satisfaction | | | |
	Excellent	Good	Fair	Poor
Workday coverage				
Holiday coverage				
Weekend coverage				
Availability of hotline				
Availability of standby				
Availability of pagers and call forwarding				
Availability of warnings before service interruptions				
Availability of automated escalation procedures				

FORM 4.4 Scalability of Network Management Services

Objects of scalability	Scalability			Lead-time expectations
	Good	Fair	Poor	
Organization				
Processes				
Instruments				
Services coverage				

FORM 4.5 Service Quality of Network Management

Quality indicators	Quantification of results	Average max./min. 95%
Throughput time of work orders		
Response time to requests		
Stability of throughput times		
Stability of response times		
Accessibility of services		

FORM 4.6 Changeability of Service Quality Indicators

Subindicators	Comments
What indicators can be changed	
Who is authorized to change indicators	
When can indicators be changed	

Organizational indicators

Organizational indicators include the following:

- Human resources—This indicator quantifies how many people are assigned to the various network management functional areas. More detailed structures can be easily supported; the target is the depth of functionality presented in Form 3.1.

- Ratio of manager to subject-matter expert—This indicator provides insight into the organizational structure. If the number is high, the structure is called *shallow*. If the number is low, the structure is called *deep*. External resources can also be included and surveyed.

- Size of communications network—This indicator quantifies and positions the company in terms of networking. The number of managed objects and the bandwidth miles are the most important subindicators. *Managed objects* are components with built-in agent capabilities or objects that can be managed from the outside using wraparound technologies. *Bandwidth miles* is the aggregate value of bandwidth segments that connects all locations in the WAN. Metropolitan and local areas can be added as an option. The prerequisite is an up-to-date configuration and inventory database.

Forms 4.7 through 4.9 are shown for these indicators.

FORM 4.7 Human Resources

Organization units	No. of people
Operations	
Administration	
Planning and design	
Client contact point	

FORM 4.8 Ratio of Manager to Subject-Matter Expert

Indicator	Number	Ratio
Manager		
Subject-matter expert		
External resources		

FORM 4.9 Size of Communications Network

Indicators	Voice	Data	E-mail	Image	Video
Number of managed objects					
Bandwidth miles					
Average kbps per line					
Mbps transmitted per device, per time period					
Number of lines per site					
Propagation delay on primary and alternate routes					

Indicators for specific network management processes

Indicators used for specific network management processes are the following:

- Client contact point
- Change management
- Fault tracking and monitoring
- Performance monitoring

Each of these indicators is described in the following subsections, along with the various forms used.

Client contact point. The efficiency of the client contact point is paramount, and this indicator evaluates the level of instrumentation. It is extremely important to improve the image of this contact point by offering a better and more complete instrumentation.

Forms 4.10 and 4.11 are those to be used to determine the efficiency of the client contact point.

FORM 4.10   Efficiency of Client Contact Point

	Indicators					
	Response delay to calls		Response delay to voice mail		Response daily to E-mail	
Problems	Ave.	Max.	Ave.	Max.	Ave.	Max.
Application						
Equipment						
Facility						
Database						
Environmental						
LAN						
WAN						
Metropolitan area						

FORM 4.11   Functions Supported by Client Contact Point

	Level of satisfaction			
Function	Excellent	Good	Fair	Poor
Troubles				
Changes				
Orders				
Information enquiries				

Change management. Because of moves, acquisitions, and job rotation, changes are frequent in communication networks. Indicators for change management consist of measuring the quality of change management and the distribution of change duration for priority 1 and 2 changes. Quality investigates how efficiently changes are implemented. Subindicators are the following:

- Completeness of change requests
- Ratio of successfully implemented changes to total number of changes
- Duration of typical changes
- Number of persons involved in changes
- Number of steps in the change management process
- Confirmation of priority 1 changes
- Confirmation of priority 2 changes

Forms 4.12 and 4.13 are used for these indicators.

FORM 4.12 Change Management Quality

Time period	No. of change requests	No. of successful changes	No. of changes complete
Daily			
Weekly			
Monthly			
Quarterly			
Semiannually			
Annually			

FORM 4.13 Distribution Diagram of Change Duration

Duration	Confirmation to priority 1 changes	Confirmation to priority 2 changes
15 mins.		
20 mins.		
60 mins.		
2 hrs.		
4 hrs.		
8 hrs.		
2 days		
4 days		
7 days		
14 days		
1 month		

Fault tracking and monitoring. Fault tracking and monitoring consists of the following indicators:

- Ratio of proactive to reactive fault detection—This indicator shows how efficiently proactive performance and fault management instruments can detect problems.

- Ratio of problem detection and elimination using the interactive voice response (IVR) to total number of detected faults—This indicator quantifies how many troubles can be detected using the IVR.

- Number of faults—This indicator quantifies troubles and problems for a specific period of time. It is beneficial to differentiate between the number of troubles for applications, equipment, and facilities.

- Duration of troubles—This indicator quantifies the duration of troubles. Depending on user needs, the impacts of many short interruptions are different from those caused by a few longer interruptions. In both cases, the availability value might be the same. Again, it is beneficial to differentiate between the duration of troubles for applications, equipment, and facilities.

- Quality of escalation procedures—This indicator shows how efficiently fault determination and diagnostics can be invoked.

- Quality of trouble-ticketing process—This indicator describes how efficiently trouble tickets are opened, tracked, and closed.

The forms used for each of these indicators are shown in Forms 4.14 through 4.19.

FORM 4.14 Proactive to Reactive Fault Detection Ratio

Time period	No. of faults	No. proactive	No. reactive	Ratio
Daily				
Weekly				
Monthly				
Quarterly				
Semiannually				
Annually				

FORM 4.15 Ratio of Problem Detection and Elimination Using the Interactive Voice Response (IVR) to Total Number of Detected Faults

Time period	No. of faults	No. solved by IVR	Ratio
Daily			
Weekly			
Monthly			
Quarterly			
Semiannually			
Annually			

FORM 4.16 Number of Faults

Indicators	Duration by Number
Applications	
Equipment	
Facilities	

FORM 4.17 Duration of Troubles

Indicators	Duration of troubles
Applications	
Fix codes	
Equipment	
Fix codes	
Facilities	
Fix codes	

FORM 4.18 Quality of Escalation Procedures

Indicators	Thresholds
Priority 1 problems	
Applications	
Equipment	
Facilities	
Priority 2 problems	
Applications	
Equipment	
Facilities	

Performance monitoring. Performance monitoring consists of the following indicators:

- availability of SLAs—This indicator checks the content and completeness of service-level agreements.

- network availability—This indicator quantifies network availability by measuring and time-stamping outages and fault resolution steps. Availability is calculated as follows:

$$\text{Availability} = \frac{\text{MTBF}}{(\text{MTBF} + \text{MTTR})}$$

FORM 4.19 Quality of Trouble Ticketing Process

Indicator	Review
Flexibility of where trouble tickets can be opened/closed	
Flexibility of workstations that allow the opening/closing of trouble tickets	
Completeness of fix codes	
What fix codes are supported	
Tracking features	
Availability of alarming	
Availability of filtering features	
Support of application programming interfaces	
Ratio of completely filled in trouble tickets to total number of trouble tickets	
Number of referrals	

where:

MTBF = mean time between failures

MTTR = mean time to repair, which can include MTTI (mean time to isolate) and MTTE (mean time to restore)

These times can be derived from trouble tickets, which are typically used as the basis for the calculations. The impacts of individual outages can be calculated for all components owned by the component experiencing problems by combining trouble tickets and configuration files.

- Availability of utilization data—This indicator investigates whether and what types of utilization data are available. These data are the basis for accounting and performance management.

- Quality of performance reports—This indicator covers the readability, depth, actuality, completeness, periodicity, distribution coverage, and presentation form of reports.

The forms for analyzing these indicators are shown in Forms 4.20 through 4.23.

FORM 4.20 Availability of SLAs

Items	Level of completeness
Who the contracting parties are	
What service indicators are included	
What workload and application-related indicators are included	
Modification procedures	
Duration of contracts	
Financial details	
Penalties for noncompliance	
Measurement tools accepted by contracting parties	
Performance reports accepted by contracting parties	

FORM 4.21 Network Availability

Indicators	MTBF	MTTR	MTTI	MTTE	Availability
Applications Fix codes					
Equipment Fix codes					
Facilities Fix codes					

FORM 4.22 Availability of Utilization Data

Indicators	Data avail.	Level of detail	Consolidation periodicity
Applications			
Equipment			
Facilities WAN MAN LAN			

Cost indicators

Cost indicators break down all expenditures by actual value and percentage of total cost. Form 4.24 on page 96 is used.

FORM 4.23 Quality of Performance Reports

Quality indicator reports	Readability	Depth	Actuality	Completeness	Periodicity	Coverage	Presentation
Service quality (availability, response time)							
Resource utilization (facilities, equipment)							
Load profiles (dialog, batch, E-mail, voice, video)							
Baselines (WAN, MAN, LAN)							
Alarms and thresholds (as part of preventive maintenance)							

FORM 4.24 Cost Indicators

Cost items	Actual value	% of total
Hardware		
Software		
Infrastructure		
Communications		
Human resources		

4.5 Job Descriptions for Network Management Staff

Successful network management operations require a well-educated management team with adequate skill levels. To make hiring and cross-education easier, the 37 most important profiles for the team are shown in Table 4.2. These job descriptions serve as a basis for evaluating the completeness of existing descriptions and documents on network management personnel.

4.6 Summary

This chapter has provided all the necessary tools to collect information during benchmarking. The questionnaires and the other forms can be modified by the benchmarking team or the clients if necessary. This chapter also provided data collection forms that have been used previously with clients. Because networks tend to be different, it is very unlikely that the same software or hardware monitoring device can be used in all environments. Thus, no in-depth treatment of monitoring was included. In real benchmarks, monitors can accomplish a lot by offering highly accurate information on resource utilization levels and service indicators. Benchmarking companies for processors and networks—in particular in IBM environments—frequently use monitoring devices, such as NetView Performance Monitor or RMF. The presented data collection tools can be expanded by monitors when monitors are available and can be rapidly customized for measuring the benchmark indicators. Chapter 5 shows many examples of how these forms are used by clients and the benchmarking team.

TABLE 4.2 Job Profiles of Team

Job title	Duties	External contacts	Qualifying experience and attributes
1. Network manager	Supervises and monitors quality of network management Estimates cost and resource requirements Reviews and approves processes and instruments Performs planning and scheduling of product implementation Develops, implements, and enforces procedural and security standards Evaluates performance of processes, instruments, and people and reports results to management Plans and directs acquisitions, training, and development projects	Other managers within information systems Some users Some vendors External consulting companies	Prior experience in statistics, mathematics, accounting, computer science, telecommunications, or equivalent Training in advanced practices, skills and concepts, administrative management, supervisory techniques, resource management, budgeting, and planning Excellent communication skills Excellent negotiation skills Excellent managerial skills
2. Network Administration Supervisor	Assists in evaluating service levels Assists in financial forecasting and budget preparation Oversees documentation and recording of network expenditures and chargebacks Analyzes security risks and assists in preparation of security plans Supervises inventory control	Other supervisors within network management Vendors Network manager	Knowledge for customer's business Ability for financial administration Communications skills with vendors Training in administrative management

TABLE 4.2 Job Profiles of Team (Continued)

Job title	Duties	External contacts	Qualifying experience and attributes
	Supervises problem-management procedures Establishes educational program for staff		
3. Inventory Coordinator	Manages online configuration application, including establishment of requirements Maintains network configuration Maintains vendor information Knows status of program and access methods used by system Maintains security of inventory control records Tracks delivery and installation of new equipment Implements coordination	Technical support Network operation, change, and problem coordinators Customer support desk Customers Service coordinator Vendors	Knowledge of communications facilities and offerings Some knowledge of systems programming and database structure Inventory-control skills Familiarity with conversion procedures and general project management
4. Change Coordinator	Coordinates overall planning and scheduling of changes Ensures adequate test plans for changes Coordinates implementation of all major hardware and software changes, installations, and detections Maintains current record of all planned and implemented changes	Vendor representatives Technical support Customer support desk Problem coordinator Inventory coordinator System and application programmers	Broad knowledge of hardware, software, and how a system is used Good aptitude for communication, negotiation, and coordination Good knowledge of review techniques

Follows up and ensures that changes are put into the system according to plan
Assumes responsibility for total communication network
Chairs change meetings
Prepares management reports
Provides input to inventory control
Evaluates vendors
Develops change management standards and procedures
Assesses impact of change on existing operations

Broad information system and teleprocessing background
Good aptitude in communication and coordination

5. Problem Coordinator

Ensures problems are routed to proper person or function for resolution
Monitors status of outstanding problems via open trouble tickets
Enforces priorities and schedules of problem resolution
Maintains up-to-date problem records, which contain problem descriptions, priority, and status
Schedules critical situation meetings with appropriate parties
Fulfills administrative reporting requirements
Cross-organizes resources if required

Vendor representatives
Technical support
Customer support desk
Change coordinator
System and application programmers
Network manager

TABLE 4.2 Job Profiles of Team (Continued)

Job title	Duties	External contacts	Qualifying experience and attributes
	Assumes responsibility for total communication network Provides input to experience files Provides input to what-if catalogs Evaluates security logs		
6. Order Processing and Provisioning Coordinator	Coordinates overall order processing and provisioning activities Consolidates orders Evaluates vendors Provides input to inventory control Develops order processing and provisioning standards Schedules provisioning Develops adequate test plans Assesses impact of new components Follows up and ensures that provisioning is conducted according to plans	Vendor representatives Technical support Network design and capacity planning Inventory coordinator Change coordinator	Broad knowledge of communication equipment Good aptitude for communication, negotiation, and coordination Good knowledge of review techniques
7. Network Operational Control Supervisor	Ensures staffing is adequate Ensures all documentation needed by planning groups is accurately created Ensures installation time schedules are met	Other supervisors within network management Customers Network manager	Knowledge of communications system software used and operator facilitates used to control it Knowledge of communications hardware

	Duties	Contacts	Qualifications
	Ensures required network changes and modifications are performed and properly coordinated with user, vendors, programmers, and operating personnel Ensures all problems reported by users through client contact point are satisfactorily resolved Prepares reports documenting effectiveness of operations group Establishes educational program for staff		Experience with hardware network-diagnostic aids Experience with software network-diagnostic aids Knowledge of problem-determination process Training in administrative management
8. Client Control Point Operator	Performs network supervision Implements first-level problem-determination procedures Maintains documentation to assist customer in terminal operation Logs problems Uses procedure guide for opening trouble tickets Reviews change activities log Delegates problems Determines problem area Assigns priorities Distributes information Performs additional duties when support desk activity is low Data entry for configuration and inventory	Customers Vendor representatives Problem and change coordinators of network administrator Network operation and technical support Network administrator for trouble tickets LAN administrator	Familiarity with functional applications and terminal equipment Training in personal relationships Clerical data-entry skills Problem-determination know-how Sensitivity to customers Understanding their business needs Pleasant telephone voice Language know-how

TABLE 4.2 Job Profiles of Team (Continued)

Job title	Duties	External contacts	Qualifying experience and attributes
	Summary of active problems for problem coordinator Entering of change information for change coordinator Monitoring of security Generating management and technical reports Recommends modification to procedures		
9. Network Operator	Observes ongoing operations and performance to identify problems Initiates corrective action where required within scope of knowledge and authority Interprets console messages from network software or applications programs and perform required actions Assists with network-oriented problem determination Implements backup procedures Implements bypass and recovery procedures for system/network problems Fulfills administrative-reporting requirements on network problems	Technical support staff Configuration and inventory function Problem and change coordinators of network administration Customer education and customer support desk Vendor representatives Network administration	Training in concepts of information and communication systems operations At least one year of network experience with Access methods Lines, clusters, and terminal types Service levels Distribution schedules Alert, intelligent, strives for efficiency Can execute bypass/recovery procedures Can perform authorized network alterations Understands the following Escalation procedures Operating of problem and change management

Role	Responsibilities	Interfaces	Qualifications
	Maintains communications with systems control Monitors all network activities Uses and invokes network diagnostic aids and tools Uses and provides input to database for problem and inventory control Performs second-level problem determination Performs network start-up and shutdown Creates schedules of network activities, such as testing and maintenance		Reporting requirements and procedures Communication skills Can use various tools, depending on the availability of such tools
10. Technician	Provides in-depth problem determination (third-level), as necessary Provides technical interface with vendors, as necessary Designs and maintains up-to-date problem determination, bypass, and recovery procedures Provides technical interface, as necessary, with application and system programmers Uses inventory-control database Ensures valid run procedures Assists with network configuration/reconfiguration Reads data dumps Starts and evaluates special-purpose diagnostics	Change coordinator Problem coordinator Client contact point operator Vendor technical personnel Application and system programmers Network operator Inventory coordinator LAN administrator	Several years experience with broad range of communication equipment, tools, and aids to maintain the network, including Network operation Network-control programs Access methods Generation of networks Configuration/reconfiguration procedures Good aptitude for communicating with people Can use diagnostic tools Understands vendor standards and procedures Patience in pursuing problems

TABLE 4.2 Job Profiles of Team (Continued)

Job title	Duties	External contacts	Qualifying experience and attributes
11. Network Performance Supervisor	Supervises setting objectives and time frames Assists network operational control in daily operations Decides access authorization for the network-management database Establishes educational program for staff	Other supervisors within network management Modeling coordinator Network manager Network operation	Excellent knowledge of communication networks Skill in using different types of instruments Understanding of expert systems Training in administrative management
12. Performance Analyst	Accomplishes tuning tasks Executes special measurements Designs and tests catalogued procedures Defines performance indicators Selects network performance oriented instruments Executes simple feasibility evaluations Tailors expert systems Supports access methods to customers Analyzes system and network trends Modifies thresholds and filters of network-management hardware and software Assists in third level of problem determination	Network-management database coordinator Vendors of measurement devices Service coordinator Modeling coordinator Network operator	Thorough technical knowledge of access methods, performance impacts from parameters, and networking components Creativity Patience in pursuing problems Detailed knowledge of instruments Understanding of expert systems

	Duties	Contacts	Knowledge/Skills
13. Database Analyst	Maintains database Selects database product Decides on database content and resolution Selects information sources	Network-performance analyst Modeling coordinator Service coordinator Vendors for measurement devices and databases	Thorough technical knowledge of database design and maintenance Communication skills Creativity for ad hoc problems Understanding of expert systems
14. Modeling Coordinator	Builds baseline models Verifies WAN, MAN, and LAN models Validates WAN, MAN, and LAN models Interprets modeling results Prepares input data Conducts performance testing Conducts pilot tests Evaluates ability of existing network configuration to meet current and near future needs	Modeling administrators Network performance analyst Service coordinator Vendors of modeling and emulation instruments Application-systems developers Advanced technology analyst	Understands how models and emulators work Communication skills for getting information on application load volumes and resource demand on communication recourses Project management skills
15. Security Management Supervisor	Evaluates security risks Prepares security plans Supervises security log evaluation procedures Assists in defining security violation thresholds Assists in elaborating on security plans for network management systems Establishes educational program for staff Supervises product-selection process	Other supervisors within network management Vendors Network management	Knowledge of customer's business and applications Knowledge of security impacts for larger organization Some knowledge of security management tools and techniques Superior personnel record Communication skills with vendors Training in administrative management

TABLE 4.2 Job Profiles of Team (Continued)

Job title	Duties	External contacts	Qualifying experience and attributes
16. Security Officer	Evaluates security risks Supervises security in real time Decides actions against penetrator Helps evaluate surveillance logs Helps elaborate on security plans Supervises security of network management system Manages passwords Helps select instruments	Security auditor Security analyst Customers Vendors of security-related instruments Other companies	Knowledge of customer information flows Knowledge of security impacts for larger organization Superior personnel record Ability to make decisions rapidly Some communication skills In-depth knowledge of security-related instruments
17. Security Auditor	Evaluates surveillance logs Helps estimate security risks Helps set violation thresholds Categorizes security risks Helps find right mix of physical and logical precautions Helps select instruments Writes report on how security plans are met	Security auditor Security analyst Customers Vendors of security-related instruments Other companies	Knowledge of customer information flows Knowledge of security impacts for larger organization Superior personnel record Ability to perform detailed clerical work Some communication skills In-depth knowledge of security-related instruments
18. Security Analyst	Defines monitoring and surveillance functions Evaluates and selects security management services Evaluates performance impacts of security techniques	Security officer Vendors Security auditor Other users	Superior personnel record In-depth knowledge of security management services and instruments Technical skills to customize products

Position	Duties	Relationships	Skills
	Constructs threat matrix Recommends instruments to be selected Supervises installation of instruments Customizes passwords and access authorization Programs instruments Establishes procedures for securing network management system		Some communication skills with vendors
19. Finance and Billing Supervisor	Assists in financial forecasting and budget preparation Oversees documentation and recording of network expenditures and chargebacks Assists in bill verification Integrates network accounting into corporate accounting Judges level of accounting accuracy Unifies voice and data accounting procedures Supervises product-selection process Establishes educational program for staff	Other supervisors within network management Corporate accounting Network manager	Knowledge of customer's business and applications Financial administration ability In-depth knowledge of accounting procedures Ability to communicate well with vendors Training in administrative management
20. Costing Specialist	Identifies cost components Collects information on costing items Activates monitors for data collection	Charging specialist Corporate accounting Accounting clerk	Training in business administration Ability to handle clerical work Specific knowledge of advanced costing techniques

TABLE 4.2 Job Profiles of Team (Continued)

Job title	Duties	External contacts	Qualifying experience and attributes
	Customizes monitors to specific needs of larger organization Follows corporate guidelines for costing Helps define accounting policy		Knowledge of station maintenance detailed records
21. Accounting Clerk	Helps identify cost components Evaluates accounting information Determines accounting equation Supervises periodic accounting processing Checks accounting accuracy Verifies vendor bills for voice and data	Costing specialist Charging specialist Corporate accounting Vendors	Training in business administration Ability to handle clerical work Specific knowledge of advanced accounting techniques Knowledge of station maintenance detailed records
22. Charging Specialist	Determines chargeback policy Implements and supervises chargeback policy Helps identify cost components Helps establish accounting equation Supports service-level negotiations Trains users in chargeback	Costing specialist Accounting clerk Corporate accounting Vendors Corporate internal users	Training in business administration Ability to handle clerical work Specific knowledge of advanced chargeback techniques Communication skills
23. Network Documentation Supervisor	Assists in developing documentation standards Prepares budgets Meets vendors	Other supervisors within network management Vendors Network manager	Knowledge of customer business Ability for financial administration Communication skills with vendors

Role	Responsibilities	Contacts	Skills/Training
	Supervises product selection Oversees integration of various products Establishes educational programs for staff		Training in administrative management
24. Database Administrator	Determines database contents, including mandatory and optional attributes Selects database product Populates configuration database Maintains configuration database Conducts Information exchange with databases	Database analyst Inventory coordinator Change coordinator Network Management Forum Database product vendor	Thorough technical knowledge of database design and maintenance Understanding of object-oriented and relational technologies Some communication skills
25. Cable Management Coordinator	Determines database attributes Surveys cable management products Selects product(s) Populates database of product(s) Maintains database of product(s) Conducts information exchange/integration with other documentation product(s)	Database analyst Inventory coordinator Change coordinator Cable management product vendors CAD/CAM product vendors Network installer	Thorough technical knowledge of database design and maintenance Understanding of CAD/CAM techniques Understanding of object-oriented and relational technologies Some communication skills
26. Implementation and Maintenance Supervisor	Assists in developing maintenance standards Prepares budgets Meets contractors Assesses security risks Supervises product(s)/service(s) selections Distributes responsibilities	Other supervisors within network management Vendors and contractors Network manager	Knowledge of customer business Ability for financial administration Communication skills with vendors Training in administrative management

TABLE 4.2 Job Profiles of Team (Continued)

Job title	Duties	External contacts	Qualifying experience and attributes
	Establishes educational program for staff		
27. Implementation Coordinator	Understands workorders Schedules implementation of workorders Executes implementation of workorders Documents implementation results Updates database after implementation	Planning and design Vendors Change coordinator Network operations control supervisor	In-depth knowledge of facilities and networking equipment Ability for high-quality technical work Some experience with databases and documentation systems Training in technical disciplines
28. Service Level Coordinator	Assumes responsibility for all communication networks Negotiates service levels Evaluates service-oriented parameters Provides feedback to capacity planning Designs and generates service reports Costs service levels and negotiates chargeback Assumes responsibility for security management	Capacity planning and design Operations support Performance monitoring and tuning Customers	Excellent communication skills Overview on communication networks Understands the relationship between service level, costs, and resources utilization
29. Standards Supervisor	Defines company-internal standards	Standard organizations Database administrator	Broad knowledge of network management standards

Role	Responsibilities	Contacts	Skills
	Supervises migration to standards Supervises implementation of standards	Database analyst	Some communication skills
30. Quality Assurance Officer	Reviews quality levels across disciplines Recommends quality levels and documentation Interfaces with management and users to publish quality results Administers "return on investment" records of services in relation to quality Monitors contract penalties with vendors	Vendors Users Standards bodies	Broad knowledge of quality standards Knowledge of ISO-9000 and TQM Some communication skills Basic knowledge of databases, reporting, and statistical techniques
31. Inspector	Develops inspection and test plans Conducts inspections on communication equipment, including spare parts Reports on status of components Provides input to change management	Change coordinator Inventory coordinator Vendors	In-depth knowledge of communication equipment In-depth knowledge of inspection procedures Some background on project management Basic communication skills
32. User Administration Supervisor	Distributes responsibilities to central and remote sites Selects instruments Oversees allocation of instruments to central and local sites Defines user profiles Works out processes and procedures	Other supervisors within network management Users Network manager	Knowledge of customer business Ability for financial administration Communication skills with vendors Training in administrative management

TABLE 4.2 Job Profiles of Team (Continued)

Job title	Duties	External contacts	Qualifying experience and attributes
	Keeps contacts with users Establishes educational program for staff		
33. Central User Administrator	Supervises central and local operations Registers user complaints Opens and refers trouble tickets Solves problems depending on priorities and assignments Communicates with users Communicates with vendors Generates reports Reviews reports of local user administrators Evaluates products	Client contact point Users Vendors Inventory coordinator Change coordinator Service coordinator	Strong background of user administration products Communication skills Negotiation skills
34. Local User Administrator	Supervises local operations Registers user complaints Opens and refers trouble tickets Solves problems, depending on priorities and assignments Communicates with users Generates reports Evaluates products	Client contact point Users Inventory coordinator Change coordinator Service coordinator	Strong background of user-administration products Communication skills Negotiation skills

	Responsibilities	Interacts with	Skills/Knowledge
35. Planning and Design Supervisor	Identifies events requiring response from planning and design group Develops strategy for producing required response Assigns priorities to projects to be completed Monitors progress of each project Prepares migration and installation plans Establishes education program for staff	Other supervisors within network management organization Network manager Vendors Business planners of the larger organization	In addition to managerial skills, good understanding of current environment and how changes in that environment affect network Good understanding of principles of network design and communication hardware and software Training in administrative management
36. Business Planner	Pursues business plans of the larger organization Defines and works with natural forecasting units that represent the business rather than networking indicators Characterizes and represents present networking workload Projects networking workload Develops installation and migration plans Assigns due dates for installations Supervises pilot installations Prepares network design alternatives that meet future needs Produces fallback plan	Business planners of the larger organization Network performance analysts Vendors of projection tools Modeling coordinator Application systems developers	Communication skills Knowledge of business of the larger organization Some experience in communication networks In-depth knowledge of projection techniques Political skills Planning and scheduling skills Communication skills in working with vendors Detailed knowledge of fallback procedures

TABLE 4.2 Job Profiles of Team (Continued)

Job title	Duties	External contacts	Qualifying experience and attributes
37. Technology Analyst	Prepares feasibility studies for network planning and network changes Evaluates communication forms, communication services, architectures, and protocols Makes reference visits Evaluates utilization and service levels Accomplishes communication analysis Performs hardware and software selection	Application systems developers Service-level coordinator Communication providers Vendors Inventory and assets coordinator Other users	Communication skills Overview on communication forms, services, and networks In-depth knowledge of service level

5

Sample Data Collected for Benchmarking Indicators

5.1 Introduction

Chapters 3 and 4 have provided the concepts of data collection, allowing forms, questionnaires, and indicators to be assembled by the benchmarking team and sent out to clients. The difficult part is then to receive accurate responses from the client. In this respect, the benchmarking team can face the following problems:

- The accuracy of data is not sufficient.
- Data collection takes too long.
- It is difficult to arrange meetings for interviews.
- Interview partners do not cooperate.
- Observations take too long.
- Contradictory answers are provided in sensitive areas.
- Data is not provided for certain areas.

Most likely, not all these problems converge on the benchmarking team, but some do. When reviewing the collected data that has been provided in this chapter, the reader must not forget how difficult it is to collect these data. Companies behind the data in this chapter remain anonymous.

The samples presented here are not related. Different data was provided for Forms 3.1, 3.2, 3.3, 3.4, and 3.5, as well as the answers to the interview questions and the in-depth evaluation of network management functions. Two companies were selected to collect the data for the forms. Five other companies provided input for the indicators. No correlation exists between the entries. In some instances, the data have been changed, extended, or omitted to improve the readability of the entries. The purpose of the collected data in this chapter is to indicate what type of answers the benchmarking team should look for. In certain cases, realistic data are close to those presented in this chapter. Interpretation, reporting, and comparisons follow in Chapters 6 and 7.

5.2 Preliminary Questionnaire

Using the questionnaire introduced in Chapter 4 and provided in Appendix A, the following example shows a typical response to a completed sample questionnaire. Although the form is usually sent out and returned prior to the start of benchmarking, companies typically do not send it back but submit the answers to the benchmarking team when the team starts its on-site work. Not answering certain questions or not giving necessary details is also typical. The gaps from the questionnaire can be filled in during the on-site interviews. It is, however, not recommended to start the on-site interviews until these questions have been answered. Much useful data can also be gathered with a telephone interview.

Redundancy between the various sample questionnaires is intentional; it is expected that different persons answer different questions. If the answers differ, the benchmarking team must find out the truth and the reasons for the different responses.

The responses are *intentionally* not always complete or accurate. Sometimes company and product names are *intentionally* mixed up. Everything is intended to present real-life conditions to the reader.

1. **Company profile**
    1.1 What business is the company in?
        Finance
    1.2 What are the major locations (sites) of the company with communication needs?
        New York, Los Angeles, Chicago, London, Frankfurt, Singapore, Tokyo, Sydney, Sao Paolo
    1.3 What communication needs exist today and in the future?
        Voice **Yes**   Data **Yes**   Video **Yes**
        Image **Yes**   E-Mail **Yes**
        The company must provide services for all communication forms. Priorities are with data and E-mail.
    1.4 What principal applications exist today?
        Saving accounts, checking accounts, credit cards, automated teller machines, home banking.

1.5 What are the communication needs of these applications?
Very different; bursty for file transfer and video, more balanced for data, private voice, image and E-mail.

1.6 What applications are supported by what locations (sites)?
All principal applications must be supported by all major locations.

1.7 What is the estimated traffic volume between locations (sites) for each communications form?
Backbone: T1-range
Access range: 56 kbps

2. Existing networks

2.1 Data networks

2.1.1 Is the network hierarchical—point-to-point—or a combination of backbone and access networks?
The network consists of a backbone part that connects major locations and an access part that offers connectivity to banks. Figure A.1b applies. New York is the master site with backup in the New York/New Jersey area.

2.1.2 Is the network peer-to-peer?
No

2.1.3 Is the network a combination of both logical alternatives sharing a common physical network?
No

2.1.4 How many domestic and international networks are in operation?
Domestic: 3   (SNA, X.25, and Frame Relay)
International: 1   (X.25)

2.1.5 What architectures are supported in the following networks?
WAN: **SNA**
MAN: **FDDI**
LAN: **Ethernet, token ring**

2.1.6 What protocols are in operation?
Routed protocols: **TCP/IP, IPX/SPX**
Nonrouted protocols: **SNA/SDLC**

2.1.7 What principal operating systems are in use?
Mainframes: **MVS, Unix, VM**
Servers: **Unix, OS/2**
Clients: **Unix, Windows, OS/2, DOS**
Network operating systems: **Novell**

2.2 Voice networks

2.2.1 Architecture and structure of the voice network
Star with distributed PBXs

2.2.2 How many domestic and international networks are in operation?
One

2.2.3 Is the bandwidth shared with data networks?
Yes, using multiplexers

2.3 Other networks

2.3.1 What other networks are in operation?
Video networks: **Yes, it is shared with the data and voice network. Bandwidth is provided on-demand pushing out voice if necessary.**

Cellular networks: Not yet
Satellite networks: Trial status for South America and Africa

2.3.2 What applications are supported by these networks?
All applications referenced in 1.6

2.3.3 If the company has more than one network:
Are multiple voice networks linked? does not apply
Are multiple data networks linked? yes
Are voice and data networks integrated? yes
Are multiple mail networks linked? no
Are multiple video networks linked? no
Are all the above linked to each other? yes, sharing the backbone bandwidth

2.4 What backup strategies and components are in use?
At the physical level? Yes
At the logical level? Yes
At the data backup by voice channels? Yes
At the public networks? Yes
At the ISDN facilities? No

3. Transmission facilities and data rates for networks

3.1 Domestic lines
Backbone networks: 34 T1 digital circuits
Access networks: 420 9.6 kbps analog circuits; 540 56/64 kbps digital circuits

3.2 International networks
Backbone networks: 10 T1 digital circuits
Access networks: 12 4.8 kbps and 33 9.6 kbps analog circuits; 42 56/64 kbps digital circuits

3.3 What value-added services are in use?
Packet switching: Yes
Frame relay: Yes
Electronic mail: Yes
LAN-to-LAN bridging service: No

3.4 Who are the providers of the value-added services?
AT&T, Infonet

4. Networking equipment

4.1 How many WANs are in use and who is the vendor for the equipment?

Equipment	Qty	Vendor
Multiplexers	52	Timeplex, Newbridge
Modems	appr. 800	Codex, IBM, Racal, Paradyne
Packet switches	42	Hughes
Front-end processors	28	IBM
Matrix switches	phased out	Dataswitch
Packet assembly/ disassembly	number is not available	
Satellite nodes (VSAT)		
Frame relay nodes	8	Stratacom
ATM nodes	planned	Fore Systems
Concentrators	some are in the access networks	
Protocol converters	some	

4.2 How many LANs are in use and who is the vendor for the equipment?

Equipment	Qty	Vendor
Ethernet	145	Many vendors
Token ring	98	IBM
Token bus		
Arcnet		
Routers	40	Cisco
	26	Wellfleet
Bridges	trial status, wide use is unlikely	
Extenders		
Gateways		

5. Personnel

5.1 Is the organization centralized or decentralized?

Centralized

5.2 What functions are supported?

Function	No. personnel
Management	4
Planning and design	12
Operations	74
Client contact point	22
Administration	48
Total	160

Job descriptions are confidential and cannot be disclosed.

6. Costs

6.1 Transmission and communication facilities

Domestic lines		Installation	Fixed costs	Operating costs (monthly)	Subtotal
34	T1	2 × $327			$22,236
420	9.6 analog	$207		$86,940	
540	56 digital	$207		$111,780	
34	T1			$285,600	$285,600
420	9.6 analog			$186,113.92	$186,113.92
540	56 digital			$269,578.80	$269,578.80
International networks					
5	T1 (France)	2 × $327		$3,270	
5	T1 (GB)	2 × $327		$3,270	
6	4.8 (France)				
16	9.6 (France)	22 × 4,000FF			FF88,000.00
6	4.8 (GB)				
17	9.6 (GB)	23 × 520L			L11,960
21	64 (France)	21 × 4,000FF			FF 84,000
21	64 (GB)		21 × 520.00 L		L10,920
5	T1 (France)			$158,450	$158,450
5	T1 (GB)			$207,800	$207,800
22	4.8 + 9.6 (GB)			L85,215	L85.215
23	4.8 + 9.6 (France)			FF177,320	FF177,320
21	64 (GB)			L91,980	L91,980
21	64 (France)			FF220,500	FF220,500
Total					

Value-added services in use?

Packet switching, frame relay, electronic mail

Providers of value added services?

AT&T, Infonet

6.2 Communication hardware (provide fixed costs, depreciation, operating costs, and maintenance costs for the following)

6.2.1 Wide area networks

6.2.2 Local area networks

6.3 Communication software (provide fixed costs, depreciation, operating costs, and maintenance costs for the following)

6.4 Personnel (provide salaries, training costs, travel costs, benefit costs, and overhead/occupancy costs for the functional areas)

6.5 Infrastructure (provide fixed costs, operating costs—value-added-tax, taxes, interests, etc.—and maintenance)

7. Network management

7.1 What network management functions are supported?
fault, configuration, performance, security, and accounting management

7.2 What network management instruments are in use?
Integrators: NetView/390 in mainframe from IBM
Element management systems for WANs and LANs: TimeView, Mainstreet, Comsphere, Codex 9800, Optivity, Site Manager, CiscoWorks, Lan Network Manager
Management platforms: OpenView as middle range integrator from Hewlett-Packard
WAN monitors: Wandel & Goltermann
WAN analyzers: Wandel & Goltermann
LAN monitors: Sniffer (Network General)
LAN analyzers: Sniffer (Network General)
Security management systems: RACF for host processors
Network design tools: None
Network optimization tools: None
Network management databases: nformix
Reporting software: Based on spreadsheets, both Lotus and Excel
Client contact point instruments: Automated Call Distributor

8. Outsourcing
Willingness to outsource? Yes
Which functions could be outsourced? Day-to-day operations
Who could be the outsourcer? Telecommunication providers
What management instruments would be outsourced? All, but passive monitoring should be supported
What percentage of human resources would be outsourced? Up to 80%

5.3 On-Site Interview Questionnaire

For the on-site interview, the benchmarking team works with the client in various sessions. Most likely, different persons should be involved because of the variety of questions. The best way to start is with managers and supervisors, who can lead the benchmarking team to others who can answer detailed technical questions.

It is recommended to repeat the questions to more than one person. Do not be surprised when the answers are different. The benchmarking team should weigh who is right and who is wrong; this process can be very subjective and difficult, so the experience of the benchmarking team is absolutely necessary here.

The following is a completed on-site interview questionnaire sample.

Part I. Investments and organization

1. What is the general network management investment philosophy of the firm for the short term, medium term, and long term?

   Medium term

2. What types of criteria are used to approve network management investments for payback, present value, and return on investment?

   All three

3. What are the detailed organizational diagrams at the corporate, business unit, and network management level?

   The Information Services Manager has four managers heading the following areas: network management, systems management, database management, and application management.

4. Are international and national voice, data, image, video, and E-mail networks managed by the same group?

   Yes

5. What is the mission of the Information Services Group?

   To meet service expectations of users.

6. Who is responsible for overall network management?

   Network Manager

7. Who is responsible for network management related business decisions?

   IS Manager

8. Who is responsible for strategic and tactical planning?

   Network Manager with IS Manager

9. Who is responsible for operations?

   Network Manager and supervisors

10. What is the mission of the network management group?

    To ensure operations continuously with high availability and acceptable service levels.

11. What are operational goals of the network management group?

    High availability up to end users.

12. How many personnel are directly involved in network management operations?

    100–249

13. What type of general skills do you look for in the network management group?

    Team spirit, but it is very difficult to maintain the team spirit due to rumors about outsourcing.

14. What are the basic network management functions performed by the network management group?

    Fault, configuration, performance, security, and accounting management functions.

15. What instruments are used by the network management group?
    Integrators, element management systems, monitors, analyzers, documentation tools, databases, reporting tools.
16. Who is responsible for the following functions, a specific business unit or the corporate information systems (IS) group?

Business unit	Corporate IS
Planning and design	x
Operations	x
Capacity planning	x
Setting standards	x
Strategic planning	x
Administration	x

17. How many persons from outside the IS organization are involved in decision making?
    Not more than 5 persons.
18. Is the Network Management Center considered cost or profit center?
    Cost Center
19. What is the annual budget including facilities and equipment for telecommunications?
    $5–$20M
20. What percentage of this budget is dedicated to network management?
    5–10%
21. What are application priorities for users in backup and disaster cases?
    Not yet specified. It will be done soon.
22. What is the estimated lost revenue when the following components are out of order?
    LAN: $30,000 for 1 hour
    WAN: $1,000,000 for 1 hour
    MAN: $300,000 for 1 hour
23. Have you ever built a business case for network management investments?
    Yes, but financial details cannot be disclosed.

Part II. Network Management Functions and Problems
1. Define network management as it relates to your firm.
   Keep the networks running 24 hours a day 7 days a week.
2. A generally accepted definition of network management is the deploying and coordinating of resources to plan, operate, administer, analyze, evaluate, design, and expand communication networks to meet service-level requirements. Do you agree with this definition? If not, how would you modify it?
   Yes
3. How does network management relate in your company to service and systems management?
   Networks and systems are managed separately. Migration is desired.
4. Listed below are the principal network management functions. State your agreement or disagreement with the functions and modify them if necessary.
   Client contact point
     Receiving problem reports
     Handling calls

Handling enquiries
Receiving change requests
Handling orders
Making service requests
Opening and referring trouble tickets
Closing trouble tickets

I agree with this list

Operations support
Determining problems using trouble tickets
Diagnosing problems
Taking corrective actions
Repairing and replacing
Referencing third parties
Backing up
Recovering
Logging events and performing corrective actions

I agree with this list

Fault tracking
Tracking manually reported or monitored faults
Tracking progress and escalating problems if necessary
Distributing information (including reports indicating fault avoidance)
Referring problems (with predictive capabilities)

I agree with this list. It is incorporated into Operations Support.

Change control
Managing, processing, and tracking service orders
Routing service orders
Supervising the handling of changes

I agree with this list

Planning and design
Analyzing needs
Projecting application load
Sizing resources
Authorizing and tracking changes
Raising purchase orders
Producing implementation plans
Establishing company standards
Maintaining quality assurance

I agree with this list. Change tracking is with Change Control.

Finance and billing
Asset management
Costing services
Client billing
Usage and outage collection
Calculation of client rebates
Bill verification
Software license control

I agree with this list

Implementation and maintenance
  Implementing change requests and work orders
  Maintaining resources
  Inspection
  Maintaining the configuration database
  Provisioning

I agree with this list

Fault monitoring
  Monitoring system and network for proactive problem detection
  Opening additional trouble tickets
  Referring trouble tickets

I agree with this list

Performance monitoring
  Monitoring system and network performance
  Monitoring service-level agreements
  Monitoring third-party and vendor performance
  Optimizing, modeling, and tuning
  Reporting on usage statistics and trends

I agree with this list

Security management
  Threat analysis
  Administration
  Detection
  Recovery
  Protecting the management systems

I agree with this list

Systems administration
  Software version control
  Software distribution
  Systems management
  Administration of user-definable tables
  Local and remote configuration of resources
  Management of names & addresses
  Applications management

I agree with this list. Applications and systems management are not included.

5. Rank the preceding functions from 1 to 11 as to their importance in your firm.

Client control point	2
Operations support	1
Fault tracking	4
Change control	5
Planning and design	10
Finance and billing	11
Implementation and maintenance	7
Fault monitoring	3

Performance monitoring   9
Security management   6
Systems administration   8

6. What percentage of your communications budget is dedicated to network management?
   6%–10%
7. What are the most important problems managing your networks today? List by priority and group them around processes, instruments, and people.
   Fault resolution is the highest priority.
8. What are the most important problems managing your networks in the future? List by priorities and group them around processes, instruments, and people.
   Technology changes, implementation of new applications, providing new telecommunication services to our branches, and integrated instrumentation.
9. What are the domestic network management requirements?
   High service quality, real-time performance reporting, and high provisioning speed.
10. What are the global network management requirements?
    Single point of contact, better provisioning, and end-to-end management
11. Between which countries would you like to have network management capabilities?
    United States, European Community, Australia, Eastern Europe, Africa and South America in addition to our present networks.
12. Between which countries do you have network management capabilities currently?
    All countries

Part III.   Network Management Implementation
1. How are your networks currently managed? (Check all that apply.)
   Centrally

At equipment level:	**selected equipment**
Physical layers only:	**selected facilities**
Logical layers only:	**IBM architecture**
Not managed:	**some devices**

2. Do you use a client contact point?
   Yes
3. How is your client contact point staffed?
   7 days with 24 hours
4. What instruments support the client contact point?
   PCs for entering and reviewing trouble tickets and Automated Call Distributor.
5. Which group within your company is responsible for each of the 11 business areas of network management?

Business area	**Who is responsible**
Client control point	Operations support
Operations support	Operations support
Fault tracking	Operations support
Change control	Administration

Planning and design	Planning and design
Finance and billing	Accounting
Implementation and	Implementation and
maintenance	maintenance
Fault monitoring	Operations support
Performance monitoring	Performance management
Security management	Security management
Systems administration	Administration

6. How many "full-time equivalents" (one person, full time, for a complete year, excluding training, vacation, and sickness) are involved in network management?

   101–200. Usually, one FTE can accomplish 1600 working hours a year.

7. What is the annual budget for network management?

   Confidential information; cannot be disclosed.

8. What are the salary ranges for network management personnel in each of the business areas?

Client control point:	$30,000–35,000
Operations support:	$32,000–38,000
Fault tracking:	$32,000–40,000
Implementation and	
maintenance:	$35,000–50,000
Fault monitoring:	$40,000–45,000
Performance monitoring:	$55,000–65,000
Security management:	$45,000–65,000
Systems administration:	$42,000–55,000
Change control:	$45,000–55,000
Planning and design:	$55,000–75,000
Finance and billing:	$45,000–70,000
Network manager:	$85,000–90,000

9. What type of network management training is available to your network management personnel? (Providers identified in question 10)

Provider of training	1	2	3	4	5
Client control point	x				x
Operations support	x	x			x
Fault tracking			x		
Change control	x				x
Planning and design	x		x	x	
Finance and billing	x		x	x	
Implementation and maintenance	x	x			x
Fault monitoring	x	x	x	x	x
Performance monitoring	x	x	x	x	
Security management	x		x	x	x
Systems administration	x		x	x	x

10. Who provides network management training?

    1. Equipment vendor
    2. Network services supplier
    3. Consultants
    4. Educational firms
    5. In-house
    Combination of all

11. What is the turnover rate of your network management staff?
    6%–15%
12. Describe your current approach to network management:
    Manager of managers using one focal point: **Yes with NetView/390**
    Manager of managers using multiple focal points: **No**
    Applications based using platform: **Yes**
    Combination: **Yes; NetView/390 and OpenView**
13. What specific network management tools are used by your company today?
    Integrators: **yes**
    Element management systems: **yes**
    Monitors: **yes**
    Administration tools: **yes**
    Network design and planning tool: **no**
14. What specific network management tools are planned for the future?
    Integrators: **yes**
    Element management systems: **yes**
    Monitors: **yes**
    Administration tools: **yes**
    Network design and planning tools: **yes with more emphasis; we hope for help from IBM and Hewlett Packard.**
15. Which of the tools used in network management are most effective?
    Element management systems for WANs and LANs.
16. Which of the tools used in network management are least effective?
    Trials with design tools have failed due to tariff changes.
17. Are outside parties engaged to perform any of the network management functions?
    Yes
18. If yes, which business areas are supported by what third-parties?
    Implementation and maintenance
19. What third parties would you prefer if you outsource network management functions?
    Network services providers
20. What criteria would you use to select outsourcers?
    Financial strength, experience, and state-of-the-art technology.

Part IV. Network Management Directions
 1. How important is managing a multiple-vendor environment?
    Very important
 2. Are you migrating toward an Open Systems Interconnection (OSI) environment today?
    Yes, but very slowly.
 3. Will OSI-based management be important in the future in 2–3 years, 4–6 years, or more than 6 years?
    2 to 3 years
 4. How important is the NetView product family from IBM for managing networks today?
    Very important
 5. How important will be the NetView product family from IBM for managing networks in the future?
    Very important

6. How important is Solve:Automation from Sterling Software for managing networks today?
   Not important
7. How important will be Solve:Automation from Sterling Software for managing networks in the future?
   Not important
8. How important is the Simple Network Management Protocol (SNMP) in managing networks today and tomorrow?
   Very important
9. How important is OMNIPoint in the network management product structure today and tomorrow?
   Important, but not highest priority
10. What are the preferences for attributes of network management platforms today and in the future?
    Today: SNMP, Autodiscovery, Event management, SQL
    Tomorrow: SNMPv2, Automapping, Event correlation
11. What network management platforms are preferred?
    OneVision. OpenView is the choice at this moment. AIX NetView.
12. What network management applications are important?
    Device and process specific
13. What level of application integration is targeted?
    Tool-bar integration and command-line integration
14. Who is responsible for integration?
    Independent software vendors

5.4   In-Depth Questions for Network Management Functions

It is very unlikely that all 63 functions of the 11 business areas of network management are evaluated in-depth. This segment offers answers for 19 functions that represent a high priority for corporations. The following chapters use this raw information and compress the information content. Usually, the completed questionnaires are provided in the appendices of the final benchmarking report.

Client contact point

Receiving problem reports.   Despite progress with monitoring techniques, a number of problems still exist with those reported electronically or over the phone by users.

1. Scope of function (what problems are expected to be received by client contact point)
   Receive problems about systems, networking components, databases and applications.
2. Time limitations of function
   This function is offered 7 days a week and 24 hours a day.
3. Frequency of executing function driven by callers and periodic polling of certain regular callers
   Driven by callers, practically no proactive monitoring or polling.

4. Person in charge of execution, support, and advice for function
   Execution is by the Help Desk. Support is by networks and systems operations. Advice can come from analysts.
5. Number of persons involved in execution, support, and advice
   Execution: 25 persons in 3 shifts
   Support: 10
   Advice: 2
6. Documentation forms used to support function
   No special forms provided by Help Desk to be filled in by the users.
7. Instruments used to support function
   Manual registration, occasional recording, but no use of interactive voice response
8. Level of automation with function
   Very low
9. Decision-making criteria used
   Depending on nature of problems, decision is made for referral.
10. Priorities used within function for individual task
    Highest priority is to understand symptoms.
11. Source of information for function
    User
12. Quality of information received for function
    Usually low, many additional questions are expected to be asked about symptoms. Answers are usually incomplete and inaccurate.
13. Destination of sending information from function
    Information is registered in trouble tickets and sent to operational support.
14. Quality control of outgoing information
    None

Handling calls

1. Scope of function
   Answering user calls who are reporting problems.
2. Time limitations of function
   This function is offered on a 7 × 24 basis. Staff tries to answer the call after the third ring.
3. Frequency of executing function
   Driven by callers
4. Person in charge of execution, support, and advice for function
   Execution: Help desk
   Support: Networks and systems operations
   Advice might come from analysts
5. Number persons involved in execution, support, and advice
   Execution: 15 persons
   Support: 3 persons
   Advice: 1 person
6. Documentation forms used to support function
   No specific forms, just the usual station maintenance detail records (SMDRs)

7. Instruments used to support function
   Manual registration, occasionally answering/recording machines, but no voice mail.
8. Level of automation with function
   Low, but occasional use of recording
9. Decision-making criteria used
   What call to take first; but no possibility of screening the caller.
10. Priorities used within function for individual tasks
    Highest priority is to identify caller. Second highest is to understand content of call and nature of problem.
11. Source of information for function
    User
12. Quality of information received for function
    Usually low because users have difficulties understanding and describing problems.
13. Destination of sending information from function
    Information is registered in trouble tickets.
14. Quality control of outgoing information
    None

Receiving change requests

1. Scope of function (what changes are considered)
   Receive change requests about system and networking components.
2. Time limitations of function
   No limitations for receiving change requests.
3. Frequency of executing function
   Once a week requests are collected, evaluated, approved, and scheduled for execution.
4. Who is in charge of execution, support, and advice for function
   Execution: Systems administration
   Support: Operations
   Advice: Planning and design
5. Number persons involved in execution, support, and advice
   Execution: 8 persons
   Support: 3 persons
   Advice: 2 persons
6. Documentation forms used to support function
   Change request forms very similar to trouble tickets.
7. Instruments used to support function
   None
8. Level of automation with function
   No automation
9. Decision-making criteria used
   Who is going to review the requests. Separating change requests by system or network.
10. Priorities used within function for individual tasks
    Three priority classes are used with this function:
       Review impact of changes

Review completeness of submitted forms
Review and consider group (mass) changes

11. Source of information for function
Requester

12. Quality of information received for function
Medium, there are usually problems because of incomplete forms.

13. Destination of sending information from function
Implementation and maintenance for execution of changes.

14. Quality control of outgoing information
None

15. Persons authorized to request changes
Practically everybody who is in charge of hardware and software for systems and networks.

16. Persons authorized to modify changes
Practically everybody who has submitted an acceptable change request.

Opening and referring trouble tickets

1. Scope of the function (what troubles are included and excluded)
All troubles are included independently whether they represent trivial troubles or subtle troubles.

2. Time limitations of function
Opening is real-time during receiving of trouble call; referral depends on the escalation procedures. Usually, trouble tickets are referred after five minutes to other areas by the Help Desk personnel.

3. Frequency of executing function
User driven; execution is usually during the trouble call.

4. Who is in charge of execution, support, and advice for function.
Execution: Help Desk
Support: Systems administration
Advice: Analysts

5. Number of persons involved in execution, support, and advice.
Execution: 25 persons (not full time)
Support: 5 persons
Advice: 2 persons

6. Documentation forms used to support function
Calls are handled individually; forms are from PNMS III from Peregrine.

7. Instruments used to support function
PNMS III from Peregrine

8. Level of automation with function
Practically all information is entered manually; no automation in copying attributes from other databases or files.

9. Decision-making criteria used
Decision making is for referrals only. Depending on the nature of troubles, they are referred to network operations, system operations, database maintenance, or application support.

10. Priorities used within function for individual tasks
   No priorities used.
11. Source of information for function
   Users who report troubles.
12. Quality of information received for function
   Low because of inaccuracies in reporting troubles.
13. Destination of sending information from function
   Referral of trouble tickets to the four principal areas: systems, networks, databases, and applications.
14. Quality control of outgoing information
   None
15. Criteria for opening trouble tickets
   All troubles require opening a ticket; if ticket already exists, priority is increased if additional users are reporting the same problem.
16. How much function can be distributed to multiple locations
   Not distributed; centrally located Help Desk opens trouble tickets.

Closing trouble tickets

1. Scope of function (criteria of closing, such as testing and confirmation of functionality)
   Trouble tickets are closed after confirmation of functionality by the user who reported the trouble.
2. Time limitations of function
   No time limitations; resolution depends on tests and confirmation of functionality by user.
3. Frequency of executing function
   Driven by trouble resolution process.
4. Person in charge of execution, support, and advice for function
   Execution: **Help Desk**
   Support: **Systems administration**
   Advice: **Analysts**
5. Number of persons involved in execution, support, and advice
   Execution: **25 persons (not full time)**
   Support: **5 persons**
   Advice: **2 persons**
6. Documentation forms used to support function
   Trouble ticket
7. Instruments used to support function
   PNMS III from Peregrine
8. Level of automation with function
   Manual closing procedure; all relevant information from the resolution process; (e.g., test results, measurements used) are attached to the trouble ticket.
9. Decision-making criteria used
   Just one is in use; whether the test and configuration data are sufficient to close the ticket.
10. Priorities used within function for individual tasks
   No priorities used.

11. Source of information for function (who gives the information to close)
    User confirmation
12. Quality of information received for function
    Good
13. Destination of sending information from function (who receives closed trouble tickets)
    Implementation and maintenance, systems administration, and use.
14. Quality control of outgoing information
    None

Operations support

Determining problems using trouble tickets

1. Scope of function (what managed objects are subject of problem determination)
   Managed objects include facilities, equipment, databases, and applications. This function identifies the problem and all the components impacted.
2. Time limitations of function
   Problem determination time limits are established on the basis of estimated impacts of managed objects:
   High-priority objects target: 10 min.
   Medium-priority objects target: 20 min.
   Low-priority objects target: 30 min.
3. Frequency of executing function
   Problem driven. Each problem received is individually addressed.
4. Person in charge of execution, support, and advice for function
   Execution: Network operations (2nd level)
   Support: Technical maintenance
   Advice: Vendors
5. Number of persons involved in execution, support, and advice
   Execution: 18
   Support: 8
   Advice: 5 (various vendors, but not full time)
6. Documentation forms used to support function
   The basis is the trouble referred from the Help Desk. In addition, documentation of the managed objects are used.
7. Instruments used to support function
   PNMS III for trouble tickets and trouble tracking. In addition, management and monitoring instruments with alarming capabilities:
   SNA-NetView/390
   LANSpy for Token Rings
   SunNet Manager for SNMP agents
   Element Management System from Timeplex and AT&T-Paradyne
   WAN/LAN monitors from Wandel & Goltermann
8. Level of automation with function
   Monitoring and alarm management are automated. Alarm correlation is manual.

9. Decision-making criteria used
   Grouping managed objects into classes requires decision-making. Following escalation steps when the time limits are exceeded also requires decision. Escalation procedures are available.
10. Priorities used within function for individual tasks
    Priorities are based on objects. Arriving and referred tickets do not impact the priorities.
11. Source of information for function
    Trouble tickets
    Management systems and monitors
    WAN components deliver status data
    LAN components polled by SunNet Manager for status data
12. Quality of information received for function
    Usually very good quality and accuracy of raw data.
13. Destination of sending information from function
    Trouble-tracking group is expected to receive information from here.
14. Quality control of outgoing information
    None

Diagnosing problems

1. Scope of function
   This function determines why the component is experiencing problems. The scope includes all managed objects, such as facilities, equipment, databases, and applications. Emphasis is, however, on facilities and equipment; diagnosis of database and application problems is heavily supported by vendors and outside groups.
2. Time limitations of function (before referral or escalation)
   Time limitations are established on the basis of estimated impacts of managed objects:
   High-priority objects target: 15 min.
   Medium-priority objects target: 30 min.
   Low-priority objects target: 45 min.
3. Frequency of executing function
   Problem driven. Each problem received is individually addressed.
4. Person in charge of execution, support, and advice for function
   Execution: Technical maintenance (3rd level)
   Support: Operations (2nd level)
   Advice: Vendors and outside consultants
5. Number of persons involved in execution, support, and advice
   Execution: 8
   Support: 5
   Advice: 3 (various part-time vendors and consultants)
6. Documentation forms used to support function
   The basis is the trouble ticket received and referred by operations, expanded by notes, measurement results, and diagnostics. In addition, documentation of the managed objects is used.
7. Instruments used to support function
   PNMS III for trouble tickets and trouble tracking. All monitoring and

management instruments identified in problem determination. The use of monitors is on-demand here, not continuous.

8. Level of automation with function
   No automation
9. Decision-making criteria used
   Decision-making includes the choice whether to switch over to spare or continue with diagnosis. Another example is group-related alarms and diagnostics.
10. Priorities used within function for individual tasks
    Priorities are based on objects. Arriving and referred tickets do not impact the priorities during diagnosis.
11. Source of information for function
    Trouble tickets, experience files, and results monitoring.
12. Quality of information received for function
    Usually very good quality and accuracy of raw data.
13. Destination of sending information from function
    Trouble-tracking group is expected to receive information from here. Occasionally, performance monitoring receives information, as well.
14. Quality control of outgoing information
    None

Fault tracking

Tracking manually reported or monitored faults

1. Scope of function (what types of faults included)
   All faults are tracked with an open trouble ticket.
2. Time limitations of function (lifecycle of trouble tickets)
   Lifecycle of trouble tickets depends on service-level agreements.
   There are three categories:

High-priority for managed objects impacting many users:	4 hrs
Medium-priority for managed objects impacting some users:	8 hrs
Low-priority for managed objects impacting few users:	24 hrs

3. Frequency of executing function
   Status review is periodic. High-priority troubles can overtake periodic status evaluation.
4. Person in charge of execution, support, and advice for function

Execution:	Systems and network administration
Support:	Operations
Advice:	Change management

5. Number of persons involved in execution, support, and advice

Execution:	3
Support:	1
Advice:	1 (part-time)

6. Documentation forms used to support function:
   Trouble tickets opened by other groups combined with fax. It is intended to combine trouble tickets with E-mail.
7. Instruments used to support function
   PNMS III in combination with fax and E-mail from Lotus.

8. Level of automation with function
   Periodic summaries (daily) of status is supported by PNMS III. Simple statistics are generated automatically.
9. Decision-making criteria used
   Built-in features highlighting missed service-level agreements requesting decisions about priority changes and escalation.
10. Priorities used within function for individual tasks
    High-priority managed objects are treated with higher priority. The tracking function does not, however, differentiate between open trouble tickets. In other words, setting priorities is manual.
11. Source of information for function
    Trouble tickets
12. Quality of information received for function
    Acceptable; occasionally, fields are missing in trouble tickets.
13. Destination of sending information from function
    Help Desk usually; sometimes information goes directly to users.
14. Quality control of outgoing information
    None
15. Polling or eventing for data collection
    No real data collection
16. Automatic changing of priorities
    It is supported by PNMS III.

Change control

Supervising the handling of changes

1. Scope of function (what changes included)
   Change requests are accepted for facilities, equipment, databases, and applications.
2. Time limitations of function
   Time limits are in use by managed objects categories. Within each group, three priorities are in use. These priorities are documented in service-level agreements.
3. Frequency of executing function
   Fixes: event-driven
   Immediate changes: event-driven
   Periodic changes: weekly
4. Person in charge of execution, support, and advice for function
   Execution: Systems and networks administration
   Support: Performance analyst
   Advice: Planning and design
5. Number of persons involved in execution, support, and advice
   Execution: 4
   Support: 1
   Advice: 1
6. Documentation forms used to support function
   For supporting the change management process, change request forms are used. The same form is used for documentation.
7. Instruments used to support function
   PNMS III from Peregrine

8. Level of automation with function
   Sorting, processing, and reporting are automated.
9. Decision-making criteria used
   Decision-making is necessary to group the incoming change requests by urgency and managed object.
10. Priorities used within function for individual tasks
    Execution priorities are set for fixes and immediate changes. Other priorities are set by managed objects.
11. Source of information for function
    Help Desk with change requests; Operations with fixes and immediate changes.
12. Quality of information received for function
    Satisfactory
13. Destination of sending information from function
    Implementation and maintenance; copy to Help Desk.
14. Quality control of outgoing information
    No special quality control steps, but the completed forms are checked for completeness.
15. Completeness of change request forms:
    Usually, 60% complete
16. Support of mass updates
    Yes, mass updates are supported.
17. Ratio of planned to implemented changes
    Estimated to be higher than 80%

Planning and design

Authorizing and tracking changes

1. Scope of function (resources considered for changes)
   Change requests are authorized for facilities, equipment, databases, and applications.
2. Time limitations of function
   Expedited changes are handled within one working day. Expedited changes include fixes and immediate changes.
3. Frequency of executing function
   On-demand for expedited changes; weekly for others.
4. Person in charge of execution, support, and advice for function

Execution:	Planning and design
Support:	Performance analysis
Advice:	Vendors

5. Number of persons involved in execution, support, and advice

Execution:	3
Support:	1
Advice:	1 (part-time)

6. Documentation forms used to support function
   Change request form
7. Instruments used to support function
   PNMS III from Peregrine
8. Level of automation with function (evaluate consequences of changes by using configuration database)

No real automation, but certain processing tasks, including sorting, reporting, and impact analysis, are supported by programs.

9. Decision-making criteria used
   A number of internal decision-making criteria are used, including the following:
   If change request forms are incomplete, reject them.
   If impacts are major, authorization takes more work.
   Look for fallback procedures if changes fail during execution.

10. Priorities used within function for individual tasks
    Expedited changes have higher priority than others.

11. Source of information for function
    Change handling function is the input.

12. Quality of information received for function
    Usually satisfactory

13. Destination of sending information from function
    Change control, Help Desk, occasionally to implementation and maintenance, and finance and billing.

14. Quality control of outgoing information
    Control of completeness

15. Is single point of contact (SPOC) used
    Yes, for all change requests submitted to the Help Desk.

16. How are contacts maintained with change control
    Contacts are good.

Establishing company standards

1. Scope of function (standards included)
   Standards include network management standards and guidelines for purchasing hardware, software, and applications.

2. Time limitations of function
   Not relevant

3. Frequency of executing function
   Periodic review twice per year. Purchase department might contact planning and design with specific questions at any time.

4. Person in charge of execution, support, and advice for function
   Execution: Planning and design
   Support: Performance analysis
   Advice: Vendors and standardization committees

5. Number of persons involved in execution, support, and advice
   Execution: 1 person
   Support: occasional
   Advice: occasional

6. Documentation forms used to support this function
   Paper-based reports and guidelines

7. Instruments used to support function
   None

8. Level of automation with function
   No automation

9. Decision-making criteria used

Limit the number of vendors and reduce the number of management protocols to a reasonable minimum.

10. Priorities used within function for individual tasks
    No priorities
11. Source of information for function
    Documentation from vendors and committees, such as the Network Management Forum, OSF, DMTF; FDDI Forum, and ATM Forum.
12. Quality of information received for function
    Sufficient
13. Destination of sending information from function
    Purchasing, finance and billing, and implementation and maintenance.
14. Quality control of outgoing information
    Review of written material, but no formal quality control.
15. How conformance to standards is controlled
    Assignments for independent consultants.
16. How are standards considered in contracts with vendors
    High priority

Finance and billing

Costing services

1. Scope of function (which equipment and facilities included)
   All equipment and facilities
2. Time limitations of function: typical duration of activity
   No time limitation to the duration of this activity.
3. Frequency of executing function:
   Quarterly
4. Person in charge of execution, support, and advice for function
   Execution: Finance and billing
   Support: Performance monitoring
   Advice: Corporate finance
5. Number of persons involved in execution, support, and advice
   Execution: 4
   Support: 1 (part-time)
   Advice: 1 (part-time)
6. Documentation forms used to support function
   Documentation is computer-based using word processors and spreadsheets.
7. Instruments used to support function
   Accounting packages, such as SMF, SAS for statistical evaluation and Excel spreadsheet for ad hoc evaluation and reporting.
8. Level of automation with function
   Automation of processing mass data and distributing information to other groups.
9. Decision-making criteria used and level of detail
   Decision is made periodically about level of detail of costing. Full responsibility costing requires resources. It has been decided to implement costing for resource groups.

10. Priorities used within function for individual tasks
    Priorities are set for WAN transmission and communication facilities, and human resources.
11. Source of information for function
    Accounting data from IBM-SMF mainframes, WANs, and DECPurchase files with maintenance records
    Personnel files with salary ranges and loaded entries
    Monitored results for LANs
    SMDR for voice-related accounting
12. Quality of information received for function
    Good, but a common denominator for all types of costs is still missing.
13. Destination of sending information from function
    Billing function
14. Quality control of outgoing information
    Review, but no formal quality control
15. Specific questions regarding finance and billing
    (Specific questions are covered with four case examples later in this chapter)

Software license control

1. Scope of function
   All operating systems, databases, and applications from all vendors.
2. Time limitations of function
   No time limitations
3. Frequency of executing function
   Sampling
4. Person in charge of execution, support, and advice for function
   Execution: Systems administration
   Support: Accounting
   Advice: Vendors
5. Number of persons involved in execution, support, and advice
   Execution: 2
   Support: 1 (part-time)
   Advice: 1 (part-time)
6. Documentation forms used to support function
   In-house forms on the basis of spreadsheets.
7. Instruments used to support function
   Excel from Microsoft
8. Level of automation with function and evaluation of counters
   No automation
9. Decision-making criteria used
   None
10. Priorities used within function for individual tasks
    No priorities set
11. Source of information for function
    Inventory data manually or semiautomatically collected.
12. Quality of information received for function
    Incomplete and usually inaccurate

13. Destination of sending information from function
    Billing function and vendors
14. Quality control of outgoing information
    None
15. Products and vendors considered
    This function is going to be re-engineered and supported by more advanced tools, probably from Microsoft.

Implemenation and maintenance

Implementing change requests and work orders

1. Scope of function (changes and work orders considered)
   Similar to the change management function, the scope includes facilities, equipment, databases, and applications.
2. Time limitations of function
   Work orders identify the time limitations for implementation.
3. Frequency of executing function
   Event-driven for fixes and immediate changes; periodic for other changes.
4. Person in charge of execution, support, and advice for function
   Execution: **Implementation group**
   Support: **Operations**
   Advice: **Performance analysis**
5. Number of persons involved in execution, support, and advice
   Execution: **3**
   Support: **1**
   Advice: **1 (part-time)**
6. Documentation forms used to support function
   Work orders that are partially paper-based, partially electronic.
7. Instruments used to support function
   Home-grown extension of PNMS III from Peregrine.
8. Level of automation with function
   No automation
9. Decision-making criteria used
   Scheduling of the implementation of work orders by priorities.
10. Priorities used within function for individual tasks and which tasks supported
    Priorities are high for fixes and immediate changes.
11. Source of information for function
    Change tracking group
12. Quality of information received for function
    Very good
13. Destination of sending information from function
    Confirmation back to the change tracking group
14. Quality control of outgoing information
    No formal quality control, but measurement and test results are enclosed.
15. Availability of procedures for mass updates after successful provisioning and changes
    No procedures for mass updates

16. Availability of testing procedures
    Yes for test procedures.

Fault monitoring

Monitoring the system and network for proactive problem detection

1. Scope of function (facilities and equipment monitored)
   Monitoring includes facilities, equipment, and systems.
2. Time limitations of function
   Once measured, evaluation and interpretation of results should be accomplished in real-time.
3. Frequency of executing function
   Event-driven (e.g., CMIP-type protocols) for WAN equipment and facilities. Periodic (e.g., SNMP-type protocols) for LANs and systems.
4. Person in charge of execution, support, and advice for function
   Execution: Performance management
   Support: Operations
   Advice: Vendor
5. Number of persons involved in execution, support, and advice
   Execution: 3
   Support: 3
   Advice: Practically every vendor (part-time)
6. Documentation forms used to support function
   Logs and consolidated measurement results. RMON is under consideration for remote LANs.
7. Instruments used to support function
   Data collection:
     WAN: NetView, TimeView
     LAN: SunNet Manager
     Systems: AIX NetView/6000
   Interpretation:
     WAN: NetView, TimeView
     LAN: SunNet Manager
     Systems: AIX NetView/6000
   Visualization:
     WAN: Graphic Monitor Facility from IBM
     LAN: SunNet Manager
     Systems: AIX NetView/6000
8. Level of automation with function:
   Data collection: highInterpretation: lowVisualization: once programs are in use from inside or outside, automation is high
9. Decision-making criteria used
   Decision has been made for hierarchical visualization, starting with the network-wide overview for both WANs and LANs. Decision has been made for filtering information in real-time arriving from WANs. LANs will work in the future with RMON for both Ethernet and Token Ring. Unix systems are measured and managed by SNMP. Mea-surement logs are kept for one working week (7 days).
10. Priorities used within function for individual tasks
    Priorities are high for backbone facilities and equipment, and for in-

terconnecting devices. Priorities are medium for LANs. Priorities are the lowest for systems.

11. Source of information for function
   All monitoring and management tools referenced earlier.
12. Quality of information received for function
   Accuracy is sufficient; indicators might be different.
13. Destination of sending information from function
   Help Desk and fault monitoring
14. Quality control of outgoing information
   None
15. Supervision of systems and networks status by indicators
   Indicators include: MTBF, MTTR, and utilization of components.
16. Capabilities of correlating alarms
   No expert systems. Use of capabilities offered by management instruments.
17. Capabilities of event notification

Events:	Yes
Alarm generation:	Yes
Alarm correlation:	Limited
Message filtering:	Yes

18. Managed objects that offer built-in monitoring capabilities

WAN elements:	modems, multiplexers, packet switches
LAN elements:	LAN segments, routers, bridges

19. Problem detection
   Status changes are visualized
   Real-time statistics are not for all indicators available
   Contact index with names, phones, faxes, E-mail addresses are available

Performance monitoring

Monitoring service-level agreements

1. Scope of function (indicators included for SLAs):
   Who are contracting parties
   Expiration date of agreement
   Procedures for modification including authorization
   Workload description and quantification
   Service-level indicators and quantification
   Determination of responsibilities
   Reporting procedures and intervals
   Accounting procedures and intervals
   Monetary charges and reimbursement for noncompliance
   Escalation procedures if deviation from agreement
   Contact names, addresses, phone, fax numbers of responsible persons
   All indicators listed.
2. Time limitations of function
   No time limitation of execution.
3. Frequency of executing function
   Periodic—every month, service indicators are recalculated and averages are computed for the last three months.

4. Person in charge of execution, support, and advice for function
   Execution: Performance management
   Support: Operations
   Advice: External consultants
5. Number of persons involved in execution, support, and advice
   Execution: 2
   Support: 1 (part-time)
   Advice: 1 (part-time)
6. Documentation forms used to support function
   Written reports are the basis of SLA-evaluation.
7. Instruments used to support function
   Monitors listed for data collection
   SAS for generating statistics, ad hoc, and permanent reports
   Some ad hoc reports can be generated by Excel
8. Level of automation with function:
   Automatic compression of collected data
   Interpretation and consolidation are manual
9. Decision-making criteria used and escalation for noncompliance
   Escalation procedures are preprogrammed and followed during execution. Escalation is on a monthly basis for each indicator. Expedited escalation for one-level higher if the same indicator has been violated three consecutive times.
10. Priorities used within function for individual tasks
    No special priorities within the task.
11. Source of information for function
    Monitored and processed data.
12. Quality of information received for function
    Good quality
13. Destination of sending information from function
    Accounting and occasional end user
14. Quality control of outgoing information
    None
15. What reports generated
    Permanent reports are:
        Average values of SLA-indicators for last month
        Changes made during last month
        Comparison for SLA indicators for last three months
        Average values of SLA indicators for last three months
        Workload summary
    Ad hoc reports:
        Details on specific SLA-indicators

Optimizing, modeling, and tuning

1. Scope of function (indicators and managed objects considered)
   Indicators considered include availability, response time, workload, and resource utilization. Managed objects considered are the parameters of the communication software (VTAM, NCP) and the operating systems of networking components (Unix and OS/2).

2. Time limitations of function
   Mission-driven; usually timeframes of 3 to 4 weeks are realistic
3. Frequency of executing function
   Function is usually mission-driven
4. Person in charge of execution, support, and advice for function
   Execution: Performance analyst
   Support: Vendors
   Advice: Vendors
5. Number of persons involved in execution, support, and advice
   Execution: 1 permanent and 2 part-time persons
   Support: Several different people, part-time
   Advice: Several different people, part-time
6. Documentation forms used to support function
   Project file and optimization handbook
7. Instruments used to support function
   All monitoring and management instruments are in use
   SAS database with performance indicators
   Best/Net is in use for the logical SNA-network
   Internetix modeling instruments are in use for Ethernet and
   token ring
8. Level of automation with function
   No automation
9. Decision-making criteria used and feasibility of hypothesis
   Decision-making is supported by go/no-go, depending on the techni-
   cal and financial feasibility of the hypothesis.
10. Priorities used within function for individual tasks
    No priorities are required within this function.
11. Source of information for function
    Monitoring and management instruments. User can help with work-
    load estimates.
12. Quality of information received for function
    Workload estimates are usually inaccurate.
13. Destination of sending information from function
    Requesters that usually include planners, designers, and end users.
14. Quality control of outgoing information
    Changes of indicators and parameters are tested before final imple
    mentation.
15. Is baselining supported?
    Baselining is not yet supported.
16. Any preference toward analytic or simulative techniques?
    No preferences, but Best/Net is analytic; Internetix LAN products are
    simulative.

Security management

Protecting the management systems

1. Scope of function (solutions for authorization and authentication)
   Authorization and authentication

2. Time limitations of function
For serious violations, actions must be taken immediately. Actions include the following:
    Deactivation of damaged configuration
    Back to last but one uncorrupted configuration
    Lock all accesses to the network management product
3. Frequency of executing function
Protection and monitoring are continuous
4. Person in charge of execution, support, and advice for function
    Execution: Security management
    Support: Systems administration
    Advice: External consultants
5. Number of persons involved in execution, support, and advice
    Execution: 1 (part-time)
    Support: 1 (part-time)
    Advice: Several persons, part-time
6. Documentation forms used to support function
Logs and written reports
7. Instruments used to support function
No special instruments, but use of Unix-based features for authorization and personalized data for authentication.
8. Level of automation with function
Detection is automatic; actions to be taken are not automated.
9. Decision-making criteria used
Decision-making criteria are based on security impacts. Config-uration violations are considered the most severe.
10. Priorities used within function for individual tasks
Configuration violations have the highest priority
11. Source of information for function
Alarms from authorization and authentication tools
12. Quality of information received for function
Satisfactory
13. Destination of sending information from function
Depending on the severity, network manager, or other managers
14. Quality control of outgoing information
None
15. Availability of more detailed description of security features of network management system
Not beyond Unix-capabilities; NetView is protected by RACF in the mainframes.
16. Partitioning capabilities
No
17. Security management solutions in place
Chip cards for authentication; passwords for authorization.

Systems administration

Software distribution
1. Scope of function
Distribution of operating systems and applications.

2. Time limitations of function
   This activity is limited to the night shift from 10:00 PM to 6:00 AM.
3. Frequency of executing function
   Push technology is used because of the large number of distributed servers and workstations.
4. Person in charge of execution, support, and advice for function
   Execution: Systems administration
   Support: Performance analysts
   Advice: Vendors
5. Number of persons involved in execution, support, and advice
   Execution: 3
   Support: 1
   Advice: Various vendors, part-time
6. Documentation forms used to support function
   Project scheduling package for scheduling and logging.
7. Instruments used to support function
   IBM distribution tools and project-scheduling package.
8. Level of automation with function
   High-level during "pushing" versions to destinations. Low-level during preparing the individual packages.
9. Decision-making criteria used; backup when distribution fails
   Decision about backup when distribution fails. If distribution fails, the last version will be reactivated.
10. Priorities used within function for individual tasks
    No priorities are necessary within this function.
11. Source of information for function
    Vendors about new versions and releases.
12. Quality of information received for function
    Satisfactory
13. Destination of sending information from function
    Finance and billing (accounting) confirming the new version or release.
14. Quality control of outgoing information
    Review of all documents
15. Combination with software licensing
    Not in use, but under consideration.

5.5 Completed Forms for Company A

The network management organization of Company A has the following structure:
Network Manager (NM)
   –Help Desk (HD)
   –Administrator (AD)
   –Technical Support (TS)
   –Network Operator (NO)
This company represents a relatively small network management organization. Prior to evaluating the forms, note that it is expected that just a sub-

set of network management functions is fully supported, few instruments
and protocols are in use, and certain job titles are not supported at all. This
example is very valuable for companies with unsophisticated networks and
businesses just building their network management organization.

The results are presented using Forms 3.1 through 3.5, which were introduced in Chapter 3. Table 5.1 is Form 3.1, Table 5.2 is Form 3.2, etc.

TABLE 5.1 Network Management Functions for Company A (Form 3.1)

	Fully supported	Partially supported	Not supported	Who is in charge
Client Contact Point				
Receiving problem reports				
Handling calls	x			HD
Handling enquiries		x		HD
Receiving change requests				
Handling orders				
Making service requests				
Opening and referring trouble tickets	x			HD
Closing trouble tickets	x			HD
Operations Support				
Determining problem using trouble tickets	x			TS
Diagnosing problem	x			TS
Taking corrective action		x		TS
Repairing and replacing		x		TS
Referring to third parties		x		TS
Backup and reconfiguration		x		TS
Recovery		x		TS
Event logging and corrective actions				
Fault Tracking				
Tracking manually reported or monitored faults				
Tracking progress and escalating problems if necessary				
Information distribution				
Referral				
Change Control				
Managing, processing, and tracking service orders				
Routing service orders				
Supervising handling of changes		x		AD

	Fully supported	Partially supported	Not supported	Who is in charge
Planning and Design				
Analyzing needs				
Projecting application load				
Sizing resources				
Authorizing and tracking changes		x		AD
Raising purchase orders				
Producing implementation plans				
Establishing company standards				
Quality assurance				
Finance and Billing				
Managing assets		x		AD
Costing services		x		AD
Billing clients				
Usage and outage collection				
Calculating client rebates				
Verifying bills				
Software license control				
Implementation and Maintenance				
Implementing change requests and work orders		x		AD
Maintaining resources				
Inspection				
Maintaining configuration database				
Provisioning				
Fault Monitoring				
Monitoring system and network for proactive problem detection				
Opening additional trouble tickets		x		NO
Referring trouble tickets		x		TS
Performance Monitoring				
Monitoring system and network performance		x		NO
Monitoring service-level agreements				
Monitoring third-party and vendor performance				
Optimizing, modeling, and tuning				
Reporting on usage statistics and trends		x		TS
Security Management				
Threat analysis				
Administration				

TABLE 5.1 Network Management Functions for Company A (Form 3.1) (Continued)

	Fully supported	Partially supported	Not supported	Who is in charge
Detection		x		AD
Recovery				
Protecting management systems				
Systems Administration				
Software version control				
Software distribution				
Systems management		x		AD
Administering user-definable tables				
Local and remote configuring resouces				
Name and address management				
Applications management				

TABLE 5.2 Use of Network Management Instruments for Company A (Form 3.2)

	In use	Planned	Owned by	Support of functional areas
Integrators				
Manager of managers				
Management platforms				
Element Management Systems for WANs				
Modems	x		TS	CCP, OS, FM, FT
Multiplexers		x	TS	CCP, OS, FM, FT
Packet switches				
Fast packet switches				
ATM switches				
SMDS nodes				
ISDN nodes				
Mobile communication				
Matrix switches				
Operations support systems				
Element Management Systems for LANs				
Bridges	x		TS	CCP, OS, FM, FT
Routers	x		TS	CCP, OS, FM, FT
Brouters				
Repeaters				
Extenders				
Hubs	x		TS	CCP, OS, FM, FT
Segments				
FDDI				
DQDB				
Monitors and Analyzers				
WAN monitor				
LAN monitor				

	In use	Planned	Owned by	Support of functional areas
WAN analyzer	x		TS	FM, PM
LAN analyzer	x		TS	FM, PM
Network monitor				
Software monitor	x		TS	FM, PM

Security Management Systems

Protection of systems and networks				
Protection of management systems				

Administration Instruments

Documentation systems		x	AD, NM	IM, SA, PD, FT, CCP
Modeling instruments				
Presentation tools				
Report generators				
Trouble-tracking tools	x		AD	OS, CCP, FT
Software distribution tools				
Software licensing tools				

Database Tools

Databases	x		AD	SA
MIB browser				
Inquiry tools				

Client Contact Point Instruments

Prediagnosis by phone
Automated call distributor
Voice mail
E-mail
Pager
Console emulator
Expert system

TABLE 5.3 Network Management Protocols for Company A (Form 3.3)

Network Management Protocols	Established	Planned	Other protocols
Proprietary protocols			
NMVT	x		
MSU		x	
Dec			
Novell	x		
Others			
CMIP			
SNMPv1	x		
SNMPv2		x	
CMOL			
CMOT			
XMP			

TABLE 5.3 Network Management Protocols for Company A (Form 3.3) (Continued)

Proprietary protocols	Established	Planned	Other protocols
RMON		x	
OMNIPoint			
Edge			
SQL			
RPC			

TABLE 5.4 Human Resources Supporting Network Management for Company A (Form 3.4)

Job titles	Established	Planned	Other job titles
Network manager	x		NM
Network administration supervisor			
Inventory and assets coordinator			
Cable management administrator			
Change coordinator		x	NA
Problem coordinator	x		NA
Order processing and provisioning coordinator			
Network operations control supervisor	x		NM
Client contact point operator	x		HD
Network operator	x		NO
Network technician	x		TS
Network performance supervisor			
Network performance analyst			
Database analyst			
Database administrator		x	AD
Modeling coordinator			
Security management supervisor			
Security officer		x	AD
Security auditor			
Security analyst			
Finance and billing supervisor			
Costing specialist		x	NM
Accounting clerk			
Charging specialist			
Documentation supervisor			
Network maintenance supervisor			
Quality assurance officer			
Inspector			
Network implementation coordinator			
Service-level coordinator			
Supervisor of standards			
User administrator			
Design and planning supervisor			
Business planner			
Technology analyst			

TABLE 5.5 Responsibility/Skill Matrix for Network Management Areas for Company A (Form 3.5)

	Client Contact Point					
Functions	Functional knowledge	General telecom	In-depth telecom	Instrument knowledge	Personal communication	Project management
Receiving problem reports						
Handling calls	x			x	x	
Handling enquiries	x			x		
Receiving change requests						
Handling orders						
Making service requests						
Opening and referring trouble tickets	x			x		
Closing trouble tickets	x			x		

	Operations Support					
Functions	Functional knowledge	General telecom	In-depth telecom	Instrument knowledge	Personal communication	Project management
Determining problems using trouble tickets	x		x	x		
Diagnosing problems	x		x	x		
Taking corrective actions	x		x	x		
Repairing and replacing	x		x	x		
Referencing to third parties	x		x	x		
Backing up and reconfiguring	x		x	x		
Recovering	x		x	x		
Logging events and performing corrective actions						

TABLE 5.5 Responsibility/Skill Matrix for Network Management Areas for Company A (Form 3.5) (Continued)

Fault Tracking

Functions	Functional knowledge	General telecom	In-depth telecom	Instrument knowledge	Personal communication	Project management
Tracking manually reported or monitored faults						
Tracking progress and escalating problems if necessary						
Distributing information						
Referring problems						

Change Control

Functions	Functional knowledge	General telecom	In-depth telecom	Business administration	Personal communication	Project management
Managing, processing, and tracking service orders						
Routing service orders						
Supervising the handling of changes	x	x				x

Planning and Design

Functions	Functional knowledge	General telecom	In-depth telecom	Business administration	Personal communication	Project management
Analyzing needs						
Projecting application load						
Sizing resources						
Authorizing and tracking changes	x	x		x		

Raising purchase orders
Producing implementation plans
Establishing company standards
Maintaining quality assurance

Finance and Billing

Functions	Functional knowledge	General telecom	In-depth telecom	Business administration	In-depth tariff information	Project management
Asset management	x	x				
Costing services	x	x		x		
Client billing						
Usage and outage collection						
Calculation of rebates to clients						
Verification of bills						
Software license control						

Implementation and Maintenance

Functions	Functional knowledge	General telecom	In-depth telecom	Practical experiences	Personal communication	Project management
Implementing change requests and work orders	x		x	x		
Maintaining resources						
Inspection						
Maintaining configuration database	x					
Provisioning			x	x		

TABLE 5.5 Responsibility/Skill Matrix for Network Management Areas for Company A (Form 3.5) (Continued)

Fault Monitoring

Functions	Functional knowledge	General telecom	In-depth telecom	Instruments knowledge	Personal communication	Project management
Monitoring system and network for proactive problem detection						
Opening additional trouble tickets	x	x		x		
Referring trouble tickets	x	x		x		

Performance Monitoring

Functions	Functional knowledge	General telecom	In-depth telecom	In-depth instruments knowledge	Creativity and patience	Project management
Monitoring system and network performance	x	x				
Monitoring service-level agreements						
Monitoring third-party and vendor performance						
Optimizing, modeling, and tuning						
Reporting on usage statistics and trends	x	x				

Security Management

Functions	Functional knowledge	General telecom	In-depth telecom	Security issues	Creativity	Patience
Threat analysis						
Administration	x	x				x
Detection						
Recovery						
Protecting the management systems						

Systems Administration

Functions	Functional knowledge	General telecom	In-depth telecom	Instruments knowledge	Personal communication	Project management
Software version control						
Distribution of software	x	x				x
Systems management						
Administration of user-definable tables						
Configuration of local and remote resources						
Management of names and addresses						
Applications management						

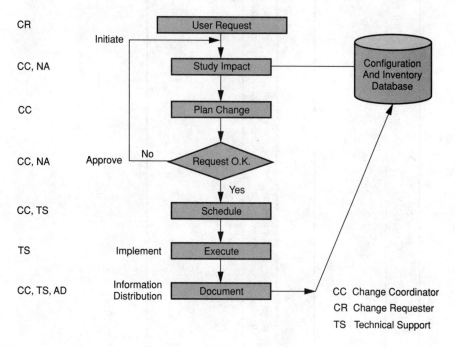

Figure 5.1 Status of the change management process

5.6 Completed Forms for Company B

The network management organization of Company B has the following structure:

Network Manager (NM)
    –Change Coordinator (CC)
    –Problem Coordinator (PC)
    –Help Desk (HD)
    –Network Operator (NO)

charge

D, NO, NA   Network Status and Supervision

D

D, NO   Dynamic Trouble Tracking

D, NA

Back-Up And Reconfiguration

A, TS

A, TS

A, TS   Diagnose and Repair

, NA

, NA

, NA, AD   Network Recovery

D, HD

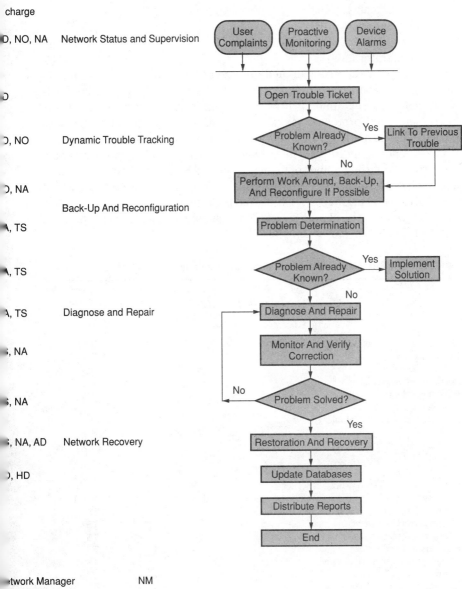

twork Manager   NM

hange Coordinator   CC
roblem Coordinator   PC
elp Desk   HD
etwork Operator   NO
etwork Analyst   NA
echnical Support   TS
ecurity Officer   SO
dministrator   AD

ure 5.2 Status of the fault management process

–Network Analyst (NA)
–Technical Support (TS)
–Security Officer (SO)
–Administrator (AD)

This company represents a medium-sized network management organization. Before evaluating the forms, note that it is expected that many network management functions are fully or partially supported, many instruments and protocols are used, and many of the job titles are supported. This example is valuable for companies with considerable network management expenditures. The networks themselves might not be very sophisticated, but the number of managed objects is expected to be considerably high. The results are presented using Forms 3.1 through 3.5, which were introduced in Chapter 3. Table 5.6 is Form 3.1, Table 5.7 is Form 3.2, etc. The status of the charge and fault management processes are shown in Figs. 5.1 and 5.2 on pages 158 and 159.

TABLE 5.6 Network Management Functions for Company B (Form 3.1)

	Fully supported	Partially supported	Not supported	Responsible party
Client Contact Point				
Receiving problem reports	x			HD
Handling calls	x			HD
Enquiry handling		x		HD
Receiving change requests				
Handling orders				
Making service requests				
Opening and referring trouble tickets	x			HD, PC
Closing trouble tickets	x			HD, PC
Operations Support				
Problem determination using trouble tickets	x			NO, TS
Problem diagnosis	x			TS
Taking corrective actions	x			TS
Repair and replacement	x			TS
Referencing third parties	x			TS
Backing up and reconfiguration		x		TS
Recovery		x		TS
Logging events and corrective actions		x		NO, TS
Fault Tracking				
Tracking manually reported or monitored faults	x			NO, TS, NA
Tracking the progress and escalation of problems if necessary		x		NO, TS, NA
Distribution of information		x		AD
Referring problems	x			TS

	Fully supported	Partially supported	Not supported	Responsible party
Change Control				
Managing, processing, and tracking service orders		x		CC
Routing service orders		x		CC
Supervising the handling of changes	x			CC
Planning and Design				
Analyzing needs		x		NA
Projecting application loads		x		NA
Sizing resources				
Authorizing and tracking changes		x		CC
Raising purchase orders				
Producing implementation plans				
Establishing company standards				
Maintaining quality assurance				
Finance and Billing				
Asset management		x		AD
Costing services	x			AD
Client billing		x		AD
Usage and outage collection				
Calculation of rebates to clients				
Bill verification				
Software license control		x		AD
Implementation and Maintenance				
Implementing change requests and work orders	x			TS
Maintaining resources		x		TS
Inspection		x		TS
Maintaining configuration database		x		TS, NA
Provisioning		x		TS
Fault Monitoring				
Monitoring system and network for proactive problem detection		x		NO, NA, TS
Opening additional trouble tickets	x			NO
Referring trouble tickets	x			TS, HD
Performance Monitoring				
Monitoring system and network performance	x			NA, TS
Monitoring service-level agreements				
Monitoring third-party and vendor performance				
Optimizing, modeling, and tuning		x		NA, TS
Reporting on usage statistics and trends to management and users	x			NA, TS

TABLE 5.6 Network Management Functions for Company B (Form 3.1) (Continued)

	Fully supported	Partially supported	Not supported	Responsible party
Security Management				
Threat analysis				
Administration		x		SO
Detection	x			SO
Recovery	x			SO
Protecting the management system		x		NA
Systems Administration				
Software version control				
Software distribution		x		AD, NA
Systems management		x		AD, NA
Administration of user-definable tables				
Local and remote configuration of resources				
Management of names and addresses		x		AD
Applications management		x		AD

TABLE 5.7 Use of Network Management Instruments for Company B (Form 3.2)

Instruments	In use	Planned	Owner	Support of functional areas
Integrators				
Manager of managers	x		NA, NM, NO	CCP, OS, FM, FT
Management platforms		x	NA, NM, NO	CCP, OS, FM, FT
Element Management Systems for WANs				
Modems	x		TS, NO	CCP, OS, FM, FT
Multiplexers	x		TS, NO	CCP, OS, FM, FT
Packet switches				
Fast packet switches				
ATM switches				
SMDS nodes				
ISDN nodes				
Mobile communication				
Matrix switches	x		TS, NO	CCP, OS, FM, FT
Operations support systems				
Element Management Systems for LANs				
Bridges	x		TS, NO	CCP, OS, FM, FT
Routers	x		TS, NO	CCP, OS, FM, FT
Brouters				
Repeaters				

Instruments	In use	Planned	Owner	Support of functional areas
Extenders				
Hubs	x		TS, NO	CCP, OS, FM, FT
Segments				
FDDI				
DQDB				

Monitors and Analyzers				
WAN monitor	x		TS, NA	FM, PM
LAN monitor	x		TS, NA	FM, PM
WAN analyzer	x		TS, NO	FM, PM
LAN analyzer	x		TS, NO	FM, PM
Network monitor				
Software monitor	x		NA	FM, PM

Security Management Systems				
Protection of systems and networks	x		SO	SM
Protection of management systems		x	NA	SM

Administration Instruments				
Documentation systems	x		AD, NM	IM, SA, PD, FT, CCP
Modeling instruments		x	NA	PD, PM
Presentation tools		x	NA	FT, FM, PM
Report generators	x		NA	PM, FB, PD
Trouble-tracking tools	x		AD, NO, HD	OS, CCP, FT
Software distribution tools		x	AD	SA
Software licensing tools		x	AD	FB

Database Tools				
Databases	x		NA	SA
MIB browser				
Inquiry tools				

Client Contact Point Instruments				
Prediagnosis by phone				
Automated call distributor				
Voice mail				
E-mail				
Pager	x		TS	CCP, OS, FM
Console emulator				
Expert system				

TABLE 5.8 Network Management Protocols for Company B (Form 3.3)

Network Management Protocols	Established	Planned	Others
Proprietary protocols			
NMVT	x		
MSU		x	
Dec			
Novell	x		
Others			
CMIP			
SNMPv1	x		
SNMPv2		x	
CMOL		x	
CMOT			
XMP			
RMON		x	
OMNIPoint			
Edge			
SQL			
RPC			

TABLE 5.9 Human Resources Supporting Network Management for Company B (Form 3.4)

Job titles	Established	Planned	Other job titles
Network manager	x		NM
Network administration supervisor			
Inventory and assets coordinator		x	AD
Cable management administrator		x	AD
Change coordinator	x		CC
Problem coordinator	x		PC
Order processing and provisioning coordinator		x	AD
Network operations control supervisor	x		NM
Client contact point operator	x		HD
Network operator	x		NO
Network technician	x		TS
Network performance supervisor			
Network performance analyst	x		NA
Database analyst			
Database administrator	x		AD
Modeling coordinator		x	NA
Security management supervisor		x	SO
Security officer	x		SO
Security auditor			
Security analyst			

Job titles	Established	Planned	Other job titles
Finance and billing supervisor			
Costing specialist	x		AD
Accounting clerk		x	AD
Charging specialist			
Documentation supervisor			
Network maintenance supervisor			
Quality assurance officer			
Inspector			
Network implementation coordinator			
Service-level coordinator		x	AD
Supervisor of standards			
User administrator			
Design and planning supervisor			
Business planner		x	NM
Technology analyst	x		NA

5.7 Indicators Provided by Five Companies

Based on the indicator definition in Chapter 4, five companies submitted indicator values to the benchmarking team. Some of the indicators could have been calculated on the basis of logs, observations, trouble tickets, or monitoring results. In such a case, however, the preparation work requires considerably more time. This segment shows selected examples for the indicators. To compare the results, each of the five companies has measured and processed results for exactly the same indicators.

Indicator values for Company 1

Indicator values for Company 1 are shown in Tables 5.11 through 5.17. Table 5.11 shows the generic indicators, Table 5.12 shows organizational indicators, and Tables 5.13 and 5.14 show indicators for specific network management processes. Table 5.13 shows the client contact point, and Table 5.14 shows change management. Table 5.15 shows fault tracking and monitoring, and Table 5.16 shows performance monitoring. Finally, Table 5.17 shows cost indicators.

TABLE 5.10 Responsibility/Skill Matrix for Network Management Areas for Company B (Form 3.5)

Client Contact Point

Functions	Functional knowledge	General telecom	In-depth telecom	Instrument knowledge	Personal communication	Project management
Receiving problem reports	x					
Handling calls	x					
Handling enquiries	x			x		
Receiving change requests				x		
Handling orders						
Making service requests						
Opening and referring trouble tickets	x	x		x		
Closing trouble tickets	x	x		x		

Operations Support

Functions	Functional knowledge	General telecom	In-depth telecom	Instrument knowledge	Personal communication	Project management
Determining problems (using trouble tickets)	x		x	x		
Diagnosing problems	x		x	x		
Taking corrective action	x		x	x	x	
Repairing and replacing	x		x	x		
Referencing third parties	x		x	x		
Backing up and reconfiguration	x		x	x		
Recovery	x		x	x	x	
Logging events and taking corrective actions	x	x		x		

Fault Tracking

Functions	Functional knowledge	General telecom	In-depth telecom	Instrument knowledge	Personal communication	Project management
Tracking manually reported or monitored faults	x	x				
Tracking progress and escalating problems if necessary				x		
Distributing information	x	x			x	
Referral	x		x			

Change Control

Functions	Functional knowledge	General telecom	In-depth telecom	Business administration	Personal communication	Project management
Managing, processing, and tracking service orders	x	x				
Routing service orders	x	x				
Supervising the handling of changes	x				x	x

Planning and Design

Functions	Functional knowledge	General telecom	In-depth telecom	Business administration	Personal communication	Project management
Analyzing needs	x	x				
Projecting application load	x	x				
Sizing resources						
Authorizing and tracking changes	x	x		x		
Raising purchase orders						
Producing implementation plans						
Establishing company standards						
Maintaining quality assurance						

TABLE 5.10 Responsibility/Skill Matrix for Network Management Areas for Company B (Form 3.5) (Continued)

Finance and Billing

Functions	Functional knowledge	General telecom	In-depth telecom	Business administration	In-depth tariff information	Project management
Asset management	x					
Costing services	x					
Clients billing	x			x		x
Usage and outage information collection						
Calculation of rebates to clients				x		
Bill verification						
Software license control						

Implementation and Maintenance

Functions	Functional knowledge	General telecom	In-depth telecom	Practical experiences	Personal communication	Project management
Implementing change requests and work orders	x		x	x		
Maintaining resources	x		x			
Performing inspections	x		x	x		
Maintaining configuration database	x					
Conducting provisioning	x		x	x		

Fault Monitoring

Functions	Functional knowledge	General telecom	In-depth telecom	Instruments knowledge	Personal communication	Project management
Monitoring system and network for proactive problem detection	x		x	x		

	Functional knowledge	General telecom	In-depth telecom	In-depth instruments knowledge	Creativity and patience	Project management
Opening additional trouble tickets	x			x		x
Referring trouble tickets	x			x		x

Performance Monitoring

Functions	Functional knowledge	General telecom	In-depth telecom	In-depth instruments knowledge	Creativity and patience	Project management
Monitoring system and network performance	x	x				
Monitoring service-level agreements						
Monitoring third-party and vendor performance						
Optimizing, modeling, and tuning	x		x	x	x	
Reporting on usage statistics and trends	x	x				

Security Management

Functions	Functional knowledge	General telecom	In-depth telecom	Security issues	Creativity	Patience
Threat analysis						
Administration	x			x		x
Detection	x		x			
Recovery	x		x			
Protecting management system	x		x		x	

TABLE 5.10 Responsibility/Skill Matrix for Network Management Areas for Company B (Form 3.5) (Continued)

Functions	Systems Administration					
	Functional knowledge	General telecom	In-depth telecom	Instruments knowledge	Personal communication	Project management
Software version control						
Distribution of software	x	x				x
Management of systems	x			x		
Administration of user-definable tables						
Configuration of local and remote resources						
Name and address management	x	x				
Application management	x				x	

TABLE 5.11 Generic Indicators for Company 1

| | Attended Operations | | | |
Indicators	Outstanding	Good	Fair	Poor
Workday coverage	x			
Holiday coverage		x		
Weekend coverage		x		
Availability of hot line	x			
Availability of standby		x		
Availability of pagers and call forwarding		x		
Availability of warnings before service interruptions				x
Availability of automated escalation procedures			x	

TABLE 5.12 Organizational Indicators for Company 1

| | Human Resources |
Organizational units	No. people
Operations	60
Administration	35
Planning and design	15
Client contact point	15
External	18

| | Ratio of Manager to Subject-Matter Expert | | |
Indicators	No.	Ratio 1	Ratio 2
Manager	15		
Subject-matter experts	110	14%	
Subtotal	125		
External	18		12%
Total	143		

TABLE 5.13 Network Management Process Indicators for Company 1
(Client Contact Point)

| | Client Contact Point Efficiency | | | | | |
| | Response delay to calls | | Response delay to voice mail | | Response delay to E-mail | |
Functions	(auto.)	(man.)	(auto.)	(man.)	(auto.)	(man.)
Handling troubles	5 secs.	30 secs.	5 mins.	15 mins.	5 mins.	30 mins.
Handling changes	10 secs.	40 secs.	10 mins.	30 mins.	15 mins.	4 hrs.
Handling orders	30 secs.	2 mins.	15 mins.	60 mins.	15 mins.	4 hrs.
Handling information enquiries	30 secs.	5 mins.	15 mins.	2 hrs.	30 mins.	8 hrs.

TABLE 5.13 Network Management Process Indicators for Company 1 (Client Contact Point) (Continued)

Client Contact Point Grade of Service

Indicators	Outstanding	Good	Fair	Poor
Application problems			x	
Database problems			x	
Equipment problems		x		
Facility problems	x			
LAN problems			x	
WAN problems	x			
MAN problems			x	
Environmental problems				x

TABLE 5.14 Network Management Process Indicators for Company 1 (Change Management)

Quality of Change Management

Time period	No. change requests	No. form complete	% complete	No. successful	% successful
Daily	25	20	80	22	88
Weekly	300	270	90	290	97
Monthly	1,400	1,350	96	1,380	98
Quarterly	4,500	4,410	98	4,460	99
Semiannually	9,600	9,500	99	9,530	99
Annually	19,400	18,800	97	19,110	98

Distribution Diagram of Change Duration

Duration	Servere priority 1 changes No.	%	Important priority 2 changes No.	%	Total	%
15 min.	5	0.83	10	1.25	15	1.07
60 min.	20	3.33	30	3.75	50	3.57
2 hr.	40	6.66	50	6.25	90	6.43
4 hr.	80	13.33	100	12.50	180	12.86
8 hr.	140	23.33	200	25.00	340	24.29
2 days	160	26.66	240	30.00	400	28.57
4 days	50	8.33	80	10.00	130	9.29
7 days	50	8.33	60	7.50	110	7.86
14 days	30	5.00	20	2.50	50	3.57
1 month	25	4.16	10	1.25	35	2.50
Total	600		800		1,400	100.00

Change Management Process Analysis

Indicators	Priorities	
	Priority 1	Priority 2
Number of persons involved in changes	7	8
Number of steps in change management process	14	18
Number of automated steps	3	4
Number of manual steps	9	10
Number of semiautomated steps	2	4

TABLE 5.15 Fault Tracking and Monitoring Indicators for Company 1

Ratio of Proactive/Reactive Fault Detection

Time period	No. faults	No. proactive	No. reactive	% radio proactive
Daily	200	50	150	25.00
Weekly	1,100	260	840	23.60
Monthly	4,500	1,000	3,500	22.20
Quarterly	14,000	3,150	10,850	22.50
Annually	60,500	16,460	44,040	27.20

Number of Troubles

Resources	Busy hour	Day	Week	Month	Quarter
Applications	25	100			
Batch		1			
Transactions		75			
Mail		24			
Equipment	10	40			
Hardware		25			
Software		15			
Facilities	15	60			
Primary circuits		58			
Backup circuits		2			
Backbone circuits		15			
Tail circuits		45			

Duration of Troubles

Resources	Total	15 mins.	1 hr.	4 hrs.	8 hrs.	1 week	4 weeks	More
Applications	100							
Batch	1	1						
Transactions	75	5	30	20	10	15	5	

TABLE 5.15 Fault Tracking and Monitoring Indicators for Company 1 (Continued)

Resources	Total	15 mins.	1 hr.	4 hrs.	8 hrs.	1 week	4 weeks	More
Mail	24	10	10	4				
Equipment	40							
Hardware	25	3	10	5	6	1		
Software	15	3	8	3	1			
Facilities	60							
Primary circuits	58	8	20	20	5	5		
Backup circuits	2				1	1		
Backbone circuits	15	2	5	5	3			
Tail circuits	45	5	15	25				

Trouble Referrals

Date (time period)	No. TT	No. filled in correctly	% ratio	1	2	3	4	more
May 6, 1994	180	160	88.89	120	40	10	5	5
March 1, 1994	190	180	94.74	125	42	13	5	5
March 24, 1994	210	205	97.62	160	30	10	5	5
August 21, 1994	170	205	95.35	150	50	5	5	5
January 29, 1994	170	150	88.24	105	30	20	5	10
May 25, 1994	200	166	93.00	170	20	10		

Fault History Time Stamps

Indicators	Detected	Service restored	Vendor notified	Isolated	Repaired
Applications					
A	8:05	8:18	8:10	9:10	10:05
B	9:15	9:20	9:20	9:25	9:30
C					
Equipment					
A	10:10	10:25	10:25	11:55	13:55
B	11:15	12:00	12:30	15:30	19:10
C					
Facilities					
A	7:00	7:05	7:00	11:00	14:00
B	9:15	9:25	9:20	11:50	12:50
C					

TABLE 5.16 Performance Monitoring Indicators for Company 1

Indicators	Availability				
	MTBF (hrs.)	MTTR (hrs.)	MTTI (hrs.)	MTTE (hrs.)	% avail.
Applications					
A	300	3	1	0.5	98.52
B	500	2	1	0.5	99.30

Indicators	MTBF (hrs.)	MTTR (hrs.)	MTTI (hrs.)	MTTE (hrs.)	% avail.
C	1,000	5	2	1	99.20
Equipment					
A	2,000	4	2	0.5	99.68
B	4,000	3	2	0.5	99.86
C	3,000	6	1.5	0.8	99.72
Facilities					
A	500	8	4	1	97.47
B	800	2	0.5	0.2	99.66
C	100	1	0.8	0.5	97.75

Availability of Utilization Data

Indicators	Data avail.	Level of detail	Consolidation periodicity	Instrument Manual	Monitored
Applications					
Batch	yes	day	day	yes	no
Online	yes	hr.	hr., day	no	yes
Mail	yes	hr.	day	yes	no
Equipment					
Server	yes	min.	hr., day	yes	no
Clients	no				
Routers	yes	min.	hr., day	yes	no
Multiplexer	yes	hr.	hr., day	yes	no
Facilities					
WAN	yes	min.	min., hr., day	yes	no
MAN	no				
LAN	yes	sec.	min., hr., day	yes	no

TABLE 5.17 Cost Indicators for Company 1

Cost items	Actual value (k)	Subtotal (k)	% of total
Hardware		100	11 11
WAN	35		
LAN	15		
Servers	30		
Clients	20		
Software		150	16.66
WAN	30		
LAN	30		
Servers	30		
Clients	10		
Application	50		
Infrastructure		50	5.55
Cabling	25		
Hubs	5		
Power supply	5		
Security surveillance	5		

TABLE 5.17 Cost Indicators for Company 1 (Continued)

Cost items	Actual value (k)	Subtotal (k)	% of total
Backup components	10		
Communications		200	22.22
Domestic lines	30		
International lines	150		
Value-added services	20		
Human resources		400	44.44
Management	50		
Operations	200		
Administration	100		
Planning and design	50		
Total		900	100.00

Indicator values for Company 2

Indicator values for Company 2 are shown in Tables 5.18 through 5.24. Table 5.18 shows the generic indicators, Table 5.19 shows organizational indicators, and Tables 5.20 and 5.21 show indicators for specific network management processes. Table 5.20 shows the client contact point, and Table 5.21 shows change management. Table 5.22 shows fault tracking and monitoring, and Table 5.23 shows performance monitoring. Finally, Table 5.24 shows cost indicators.

TABLE 5.18 Generic Indicators for Company 2

	Attended Operations			
Indicators	Outstanding	Good	Fair	Poor
Workday coverage		x		
Holiday coverage			x	
Weekend coverage		x		
Availability of hot line		x		
Availability of standby		x		
Availability of pagers and call forwarding	x			
Availability of warnings before service interruptions				x
Availability of automated escalation procedures				x

TABLE 5.19 Organizational Indicators for Company 2

Human Resources	
Organizational units	No. people
Operations	72

Organizational units	No. people
Administration	43
Planning and design	12
Client contact point	20
External	22

Ratio of Manager to Subject-Matter Expert

Indicators	No.	Ratio 1	Ratio 2
Manager	17		
Subject-matter			
experts	130	12%	
Subtotal	147		
External	22		11%
Total	169		

TABLE 5.20 Network Management Process Indicators for Company 2 (Client Contact Point)

Client Contact Point Efficiency

Functions	Response delay to calls		Response delay to voice mail		Response delay to E-mail	
	Auto. (secs.)	Man. (secs.)	Auto. (mins.)	Man. (mins.)	Auto. (mins.)	Man. (hrs.)
Handling troubles	6	25	5	20	30	1
Handling changes	15	60	15	30	30	4
Handling orders	60	60	15	60	30	6
Handling information enquiries	60	60	15	60	30	8

Client Contact Point Grade of Service

Grade of service indicators	Outstanding	Good	Fair	Poor
Application problems			x	
Database problems			x	
Equipment problems		x		
Facility problems	x			
LAN problems			x	
WAN problems	x			
MAN problems			x	
Environmental problems		x		

TABLE 5.21 Network Management Process Indicators for Company 2 (Change Management)

Time period	No. change requests	No. form complete	%complete	No. successful	% succesful
		Quality of Change Management			
Daily	10	9	90	8	80
Weekly	60	58	97	55	92
Monthly	250	240	96	231	92
Quarterly	700	685	98	674	96
Semiannual	1,350	1,290	96	1,280	95
Annual	2,800	2,705	97	2,698	96

Distribution Diagram of Change Duration

Duration	Severe priority 1 changes No.	%	Important priority 2 changes No.	%	Total	%
15 mins.	2	1.80	0	0	2	0.80
60 mins.	3	2.70	2	1.43	5	2.00
2 hrs.	5	4.50	4	2.86	9	3.60
4 hrs.	35	31.80	40	28.60	75	30.00
8 hrs.	40	36.40	55	39.23	95	38.00
2 days	15	13.60	25	17.86	40	16.00
4 days	5	4.50	8	5.71	13	5.20
7 days	2	1.80	2	1.43	4	1.60
14 days	2	1.80	2	1.43	4	1.60
1 month	1	0.90	2	1.43	3	1.20
Total	110		140		250	100.00

Change Management Process Analysis

Indicators	Priorities Priority 1	Priority 2
Number of persons involved in changes	5	4
Number of steps in change management process	9	10
Number of automated steps	2	3
Number of manual steps	7	7
Number of semiautomated steps	0	0

TABLE 5.22 Fault Tracking and Monitoring Indicators for Company 2

Time period	No. faults	No. proactive	No. reactive	% ratio proactive
	Ratio of Proactive/Reactive Fault Detection			

Time period	No. faults	No. proactive	No. reactive	% ratio proactive
Daily	150	20	130	13.30
Weekly	860	110	750	12.80
Monthly	2,950	505	2,445	17.12
Quarterly	9,100	1,650	7,450	18.13
Annually	35,800	5,210	30,590	14.55

Number of Troubles

Resources	Busy hour	Day	Week	Month	Quarter
Applications	12	48			
Batch		3			
Transactions		25			
Mail		20			
Equipment	8	32			
Hardware		10			
Software		22			
Facilities	10	40			
Primary circuits		37			
Backup circuits		3			
Backbone circuits		10			
Tail circuits		30			

Duration of Troubles

Resources	Total	15 mins.	1 hr.	4 hrs.	8 hrs.	1 week	4 weeks	More
Applications	48							
Batch	3	2	1					
Transactions	25	5	5	10	5			
Mail	20		5	5	5	5		
Equipment	32							
Hardware	10	5	4	1				
Software	22	5	5	4	4	2	2	
Facilities	40							
Primary circuits	37	2	11	5	2	15	2	
Backup circuits	3		3					
Backbone circuits	10		5	5				
Tail circuits	30	2	8	4	12	2	2	

Trouble Referrals

Date (time period)	No. TT	No. filled in correctly	Ratio (%)	No. referrals 1	2	3	4	more
June 4, 1994	210	200	95.24	170	20	10	5	5
May 25, 1994	180	175	97.22	150	20	5	5	
May 19, 1994	190	180	94.74	155	5	20	5	5
May 16, 1994	200	185	92.50	160	20	10	10	
May 10, 1994	165	160	96.97	110	25	20	5	5
May 9, 1994	192	186	96.88	152	8	15	15	2

TABLE 5.22 Fault Tracking and Monitoring Indicators for Company 2 (Continued)

Fault History Time Stamps

Indicators	Detected	Service Restored	Vendor Notified	Isolated	Repaired
Applications					
A	6:00	6:15	6:05	6:10	7:15
B	12:10	13:00	12:15	12:55	14:15
C					
Equipment					
A	21:10	22:00	21:15	21:45	22:00
B	23:20	0:00	23:35	0:30	1:00
C					
Facilities					
A	6:40	7:25	6:45	7:55	8:10
B	19:30	21:15	19:45	20:05	22:15
C					

TABLE 5.23 Performance Monitoring Indicators for Company 2

Availability

Indicators	MTBF (hrs.)	MTTR (hrs.)	MTTI (hrs.)	MTTE (hrs.)	Avail. (%)
Applications					
A	250	4	1.5	1	97.47
B	320	3	2.5	0.5	98.16
C	890	6	4	1	98.78
Equipment					
A	5,000	2	1	0.5	99.93
B	4,000	2.5	0.8	0.4	99.91
C	1,500	1.5	0.9	0.3	99.82
Facilities					
A	700	7	5	1	98.18
B	800	8	5	0.5	98.34
C	650	4	3	0.4	98.72

Availability of Utilization Data

Indicators	Data Avail.	Level of details	Consolidation periodicity	Instrument Man.	Mon.
Applications					
Batch	yes	day	day	no	yes
Online	yes	hr.	hr.	no	yes
Mail	yes	hr.	day	yes	no
Equipment					
Server	no				
Clients	no				
Routers	yes	min.	hr.	no	yes

Indicators	Data Avail.	Level of details	Consolidation periodicity	Instrument Man.	Mon.
Multiplexer	yes	min.	hr.	no	yes
Facilities					
WAN	yes	min.	day	no	yes
MAN	no				
LAN	yes	sec.	hr.	no	yes

TABLE 5.24 Cost Indicators for Company 2

Cost items	Actual value (k)	Subtotal (k)	% of total
Hardware		50	11.63
WAN	10		
LAN	10		
Servers	15		
Clients	5		
Software		80	18.60
WAN	10		
LAN	15		
Servers	10		
Clients	5		
Application	40		
Infrastructure		30	6.98
Cabling	8		
Hubs	6		
Power supply	4		
Security surveillance	2		
Backup components	8		
Communications		80	18.60
Domestic lines	25		
International lines	45		
Value-added services	20		
Human resources		190	44.19
Management	25		
Operations	110		
Administration	30		
Planning and design	25		
Total		430	100.00

Indicator values for Company 3

Indicator values for Company 3 are shown in Tables 5.25 through 5.31. Table 5.25 shows the generic indicators, Table 5.26 shows organizational indicators, and Tables 5.27 and 5.28 show indicators for specific network management processes. Table 5.27 shows the client contact point, and Table 5.28 shows change management. Table 5.29 shows fault tracking and monitoring, and Table 5.30 shows performance monitoring. Finally, Table 5.31 shows cost indicators.

TABLE 5.25 Generic Indicators for Company 3

	Attended Operations			
Indicators	Outstanding	Good	Fair	Poor
Workday coverage		x		
Holiday coverage				x
Weekend coverage			x	
Availability of hot line		x		
Availability of standby		x		
Availability of pagers and call forwarding	x			
Availability of warnings before service interruptions				x
Availability of automated escalation procedures			x	

TABLE 5.26 Organizational Indicators for Company 3

Human Resources	
Organizational units	No. people
Operations	18
Administration	18
Planning and design	5
Client contact point	5
External	6

Ratio of Manager to Subject-Matter Expert			
Indicators	No.	Ratio 1	Ratio 2
Manager	6		
Subject-matter experts	40	15%	
Subtotal	46		
External	6		13%
Total	52		

TABLE 5.27 Network Management Process Indicators for Company 3 (Client Contact Point)

	Client Contact Point Efficiency					
	Response delay to calls		Response delay to voice mail		Response delay to E-mail	
Functions	Auto. (secs.)	Man.	Auto. (mins.)	Man.	Auto. (mins.)	Man.
Handling troubles	4	30 secs.	10	30 mins.	10	30 mins.
Handling changes	12	60 secs.	15	60 mins.	30	1 hrs.

Functions	Response delay to calls Auto. (secs.)	Man.	Response delay to voice mail Auto. (mins.)	Man.	Response delay to E-mail Auto. (mins.)	Man.
Handling orders	60	1 hr.	30	2 hrs.	30	4 hrs.
Handling information enquiries	60	1 hr.	15	2 hrs.	30	8 hrs.

Client Contact Point Grade of Service

Indicators	Outstanding	Good	Fair	Poor
Application problems		x		
Database problems			x	
Equipment problems	x			
Facility problems	x			
LAN problems		x		
WAN problems		x		
MAN problems			x	
Environmental problems		x		

TABLE 5.28 Network Management Process Indicators for Company 3 (Change Management)

Quality of Change Management

Time period	No. change requests	No. form complete	% complete	No. successful	% successful
Daily	15	14	93	13	86
Weekly	75	70	93	70	93
Monthly	310	300	97	300	97
Quarterly	905	888	98	895	99
Semiannually	5,200	5,100	98	5,110	98
Annually	10,100	9,880	98	9,920	98

Distribution Diagram of Change Duration

Duration	Severe priority 1 changes No.	%	Important priority 2 changes No.	%	Total	%
15 min.	4	2.66	6	3.75	10	3.23
60 min.	8	5.33	7	4.38	15	4.84
2 hr.	12	8.00	18	11.25	30	9.68
4 hr.	18	12.00	22	13.75	40	12.90
8 hr.	50	33.33	30	18.75	80	25.81
2 days	20	13.33	40	25.00	60	19.35
4 days	24	16.00	26	16.25	50	16.13
7 days	3	2.00	2	1.25	5	1.61
14 days	6	4.00	4	2.50	10	3.23

TABLE 5.28 Network Management Process Indicators for Company 3 (Change Management) (Continued)

Duration	Severe priority 1 changes		Important priority 2 changes		Total	%
	No.	%	No.	%		
1 month	5	3.33	5	3.13	10	3.23
Total	150		160		310	100,000

Change Management Process Analysis

	Priorities	
Indicators	Priority 1	Priority 2
Number of persons involved in changes	11	8
Number of steps in change management process	16	16
Number of automated steps	2	2
Number of manual steps	12	12
Number of semiautomated steps	2	2

TABLE 5.29 Fault Tracking and Monitoring Indicators for Company 3

Ratio of Proactive/Reactive Fault Detection

Time period	No. faults	No. proactive	No. reactive	% ratio proactive
Daily	100	25	75	25.00
Weekly	500	130	420	23.60
Monthly	2,250	500	1,750	22.20
Quarterly	7,500	2,200	5,300	29.33
Annually	30,250	8,350	21,900	27.60

Number of Troubles

Resources	Busy hour	Day	Week	Month	Quarter
Applications	12	50			
Batch		1			
Transactions		36			
Mail		13			
Equipment	5	20			
Hardware		12			
Software		8			
Facilities	8	30			
Primary circuits		28			
Backup circuits		2			
Backbone circuits		5			
Tail circuits		25			

Duration of Troubles

Resources	Total	15 mins.	1 hr.	4 hrs.	8 hrs.	1 week	4 weeks	More
Applications	50							
Batch	1		1					
Transactions	36	2	10	12	5	5		
Mail	13	2	4	5	2			
Equipment	20							
Hardware	12	1	5	3	3			
Software	8	1	2	3	2			
Facilities	30							
Primary circuits	28	1	6	8	8	4	1	
Backup circuits	2			1	1			
Backbone circuits	5	1	3	1				
Tail circuits	25	1	4	5	6	6	3	

Trouble Referrals

Date (time period)	No. TT	No. filled in correctly	Ratio %	No. referrals 1 2 3 4 more
May 1, 1994	110	100	90.90	100 5 2 2 1
May 2, 1994	100	95	95.00	92 6 1 1
May 3, 1994	95	90	94.74	85 5 4 1
May 4, 1994	120	110	91.67	101 8 2 7 3
May 5, 1994	115	105	91.30	102 4 4 4 1
May 6, 1994	90	88	97.78	81 2 2 2 3

Fault History Time Stamps

Indicators	Detected	Service restored	Vendor notified	Isolated	Repaired
Applications					
A	2:30	2:50	2:35	3:10	4:00
B	4:45	4:50	4:50	6:25	8:15
C					
Equipment					
A	7:10	8:10	7:30	7:50	9:00
B	11:15	13:15	11:30	14:00	16:00
C					
Facilities					
A	16:00	17:00	16:30	17:15	19:05
B	24:00	0:30	0:15	6:25	8:30
C					

TABLE 5.30 Performance Monitoring Indicators for Company 3

			Availability		
Indicators	MTBF (hrs.)	MTTR (hrs.)	MTTI (hrs.)	MTTE (hrs.)	% avail.
Applications					
A	280	4	3	2	96.99
B	510	3	2	1	98.84
C	400	4	3	2	97.80
Equipment					
A	1,500	2	1	0.5	99.77
B	2,500	2	1	0.5	99.86
C	3,800	3	2.5	0.5	99.84
Facilities					
A	700	5	4	0.5	98.66
B	900	6	5	0.5	98.74
C	300	8	6	0.5	95.39

		Availability of Utilization Data		Instrument	
Indicators	Data avail.	Level of details	Consolidation periodicity	Man.	Monitor
Applications					
Batch	yes	week	week	yes	no
Online	yes	hour	day	no	yes
Mail	yes	month	month	yes	no
Equipment					
Server	yes	minute	hour	no	yes
Clients	no				
Routers	no				
Multiplexer	yes	minute	hour	no	yes
Facilities					
WAN	yes	hour	day	yes	no
MAN	no				
LAN	yes	minute	hour	no	yes

TABLE 5.31 Cost Indicators for Company 3

Cost items	Actual value (k)	Subtotal (k)	Total %
Hardware		70	12.73
WAN	40		
LAN	10		
Servers	15		
Clients	5		
Software		90	16.37
WAN	20		
LAN	20		
Servers	30		

Cost items	Actual value (k)	Subtotal (k)	Total %
Clients	5		
Application	15		
Infrastructure		60	11.32
Cabling	25		
Hubs	5		
Power supply	3		
Security surveillance	2		
Backup components	25		
Communications		80	14.55
Domestic lines	30		
International lines	40		
Value-added services	10		
Human resources		150	45.45
Management	10		
Operations	70		
Administration	60		
Planning and design	10		
Total		550	100.00

Indicator values for Company 4

Indicator values for Company 4 are shown in Tables 5.32 through 5.38. Table 5.32 shows the generic indicators, Table 5.33 shows organizational indicators, and Tables 5.34 and 5.35 show indicators for specific network management processes. Table 5.34 shows the client contact point, and Table 5.35 shows change management. Table 5.36 shows fault tracking and monitoring, and Table 5.37 shows performance monitoring. Finally, Table 5.38 shows cost indicators.

TABLE 5.32 Generic Indicators for Company 4

	Attended Operations			
Indicators	Outstanding	Good	Fair	Poor
Workday coverage	x			
Holiday coverage			x	
Weekend coverage			x	
Availability of hot line		x		
Availability of standby		x		
Availability of pagers and call forwarding	x			
Availability of warnings before service interruptions				x
Availability of automated escalation procedures				x

**TABLE 5.33 Organizational Indicators for
Company 4**

Human Resources	
Organizational units	No. people
Operations	11
Administration	11
Planning and design	3
Client contact point	3
External	0

Ratio of Manager to Subject-Matter Expert			
Indicators	No.	Ratio 1	Ratio 2
Manager	7		
Subject-matter experts (SME)	21	33%	
Subtotal	27		
External	0		33%
Grand total	27		

**TABLE 5.34 Network Management Process Indicators for Company 4
(Client Contact Point)**

Client Contact Point Efficiency						
	Response delay to calls		Response delay to voice mail		Response delay to E-mail	
Functions	Auto.	Man.	Auto.	Man.	Auto.	Man.
Handling troubles	6 secs.	20 secs.	10 mins.	30 mins.	1 min.	5 mins.
Handling changes	15 secs.	45 secs.	1 hr.	4 hrs.	1 hr.	4 hrs.
Handling orders	1 min.	10 mins.	1 day	2 days	1 hr.	8 hrs.
Handling information enquiries	30 secs.	5 mins.	15 mins.	30 mins.	1 min.	5 mins.

Client Contact Point Grade of Service				
Indicators	Outstanding	Good	Fair	Poor
Application problems		x		
Database problems				x
Equipment problems		x		
Facility problems	x			
LAN problems		x		
WAN problems		x		
MAN problems				x
Environmental problems			x	

TABLE 5.35 Network Management Process Indicators for Company 4 (Change Management)

		Quality of Change Management			
Time period	No. change requests	No. form complete	% complete	No. successful	% successful
Daily	10	10	100	10	100
Weekly	48	45	94	46	96
Monthly	202	190	94	190	94
Quarterly	600	580	97	578	96
Semiannually	1,200	1,160	97	1,150	96
Annually	2,400	2,285	95	2,318	97

Distribution Diagram of Change Duration

	Severe priority 1 changes		Important priority 2 changes			
Duration	No.	%	No.	%	Total	%
15 min.	3	2.65	1	1.09	4	1.98
60 min.	10	8.85	5	5.49	15	7.43
2 hr.	5	4.42	5	5.49	10	4.95
4 hr.	19	16.81	13	14.29	32	15.84
8 hr.	31	27.43	27	25.27	54	26.73
2 days	19	16.81	19	20.88	38	18.81
4 days	11	9.73	10	10.99	21	10.40
7 days	6	5.30	9	9.89	15	7.43
14 days	5	4.42	3	3.30	8	3.96
1 month	4	3.54	1	1.09	5	2.48
Total	113		91		202	100.00

Change Management Process Analysis

	Priorities	
Indicators	Priority 1	Priority 2
Number of persons involved in changes	6	5
Number of steps in change management process	12	12
Number of automated steps	2	3
Number of manual steps	9	8
Number of semiautomated steps	1	1

TABLE 5.36 Fault Tracking and Monitoring Indicators for Company 4

	Ratio of Proactive/Reactive Fault Detection			
Time period	No. faults	No. proactive	No. reactive	% ratio proactive
Daily	40	5	35	12.50
Weekly	210	20	190	9.53

TABLE 5.36 Fault Tracking and Monitoring Indicators for Company 4 (Continued)

Time period	No. faults	No. proactive	No. reactive	% ratio proactive
Monthly	815	85	740	10.43
Quarterly	2,404	240	2,162	9.98
Annually	7,210	813	6,397	11.28

Number of Troubles

Resources	Busy hour	Day	Week	Month	Quarter
Applications	5	18			
Batch		2			
Transactions		10			
Mail		6			
Equipment	4	12			
Hardware		6			
Software		6			
Facilities	3	10			
Primary circuits		9			
Backup circuits		1			
Backbone circuits		5			
Tail circuits		5			

Duration of Troubles

Resources	Total	15 mins.	1 hr.	4 hrs.	8 hrs.	1 week	4 weeks	More
Applications	18							
Batch	2		1	1				
Transactions	10	1	4	4	1			
Mail	6		3	3				
Equipment	12							
Hardware	6		2	4				
Software	6	1	1	3	1			
Facilities	10							
Primary circuits	9	1	7	1				
Backup circuits	1			1				
Backbone circuits	5	1	2	1	1			
Tail circuits	5	1	2	1	1			

Trouble Referrals

Date (time period)	No. TT	No. filled in correctly	Ratio %	No. referrals 1	2	3	4	more
July 1, 1994	40	38	95.00	10	10	5	10	5
July 2, 1994	35	31	88.57	8	10	6	9	
July 3, 1994	42	40	95.24	12	10	5	10	5
July 4, 1994	10	10	100.00	8	1	1		
July 5, 1994	52	48	92.30	15	10	10	10	7
July 6, 1994	42	40	95.24	14	8	5	11	4

Fault History Time Stamps

Indicators	Detected	Service restored	Vendor notified	Isolated	Repaired
Applications					
A	15:10	16:10	15:20	16:20	17:10
B	16:40	16:55	16:50	17:50	19:20
C	2:40	3:00	6:10	6:20	6:30
Equipment					
A	19:15	21:15	19:35	20:35	23:55
B	20:25	20:55	20:30	20:50	21:15
C	1:30	6:30	2:00	7:00	9:00
Facilities					
A	1:40	9:40	3:00	12:00	15:35
B	18:20	19:40	19:15	23:15	1:30
C	4:25	5:15	4:45	10:15	11:00

TABLE 5.37 Performance Monitoring Indicators for Company 4

	Availability				
Indicators	MTBF (hrs.)	MTTR (hrs.)	MTTI (hrs.)	MTTE (hrs.)	% avail.
Applications					
A	1,100	4	2	1	99.37
B	550	2	1	0.5	99.37
C	780	2	1	0.5	99.55
Equipment					
A	5,000	3	1	0.5	99.91
B	2,500	5	4	1	99.60
C	1,000	1	0.5	0.2	99.83
Facilities					
A	800	4	2	1	99.13
B	1,050	8	3	2	99.06
C	2,000	8	4	2	99.30

Availability of Utilization Data

Indicators	Data avail.	Level of details	Consolidation periodicity	Instrument Man.	Instrument Mon.
Applications					
Batch	yes	day	day	yes	no
Online	yes	hr.	day	no	yes
Mail	yes	day	day	yes	no
Equipment					
Server	yes	hr.	day	no	yes
Clients	no				
Routers	no				
Multiplexers	no				

TABLE 5.37 Performance Monitoring Indicators for Company 4 (Continued)

Indicators	Data avail.	Level of details	Consolidation periodicity	Instrument Man.	Mon.
Facilities					
WAN	yes	day	day	yes	no
MAN	no				
LAN	yes	min.	hr.	no	yes

TABLE 5.38 Cost Indicators for Company 4

Cost items	Actual value (k)	Subtotal (k)	% of total
Hardware		30	16.13
WAN	15		
LAN	15		
Servers	10		
Clients	5		
Software		40	21.51
WAN	5		
LAN	10		
Servers	10		
Clients	5		
Application	10		
Infrastructure		25	13.44
Cabling	5		
Hubs	8		
Power supply	1		
Security surveillance	1		
Backup components	10		
Communications		35	18.82
Domestic lines	25		
International lines	5		
Value-added services	5		
Human resources		56	30.11
Management	18		
Operations	17		
Administration	15		
Planning and design	6		
Total		186	100.00

Indicator values for Company 5

Indicator values for Company 5 are shown in Tables 5.39 through 5.45. Table 5.39 shows the generic indicators, Table 5.40 shows organizational indicators, and Tables 5.41 and 5.42 show indicators for specific network management processes. Table 5.41 shows the client contact point, and Table 5.42 shows change management. Table 5.43 shows fault tracking and monitoring, and Table 5.44 shows performance monitoring. Finally, Table 5.45 shows cost indicators.

TABLE 5.39 Generic Indicators for Company 5

| | Attended Operations | | | |
Indicators	Outstanding	Good	Fair	Poor
Workday coverage		x		
Holiday coverage		x		
Weekend coverage		x		
Availability of hot line		x		
Availability of standby		x		
Availability of pagers and call forwarding		x		
Availability of warnings before service interruptions		x		
Availability of automated escalation procedures		x		

TABLE 5.40 Organizational Indicators for Company 5

| Human Resources | |
Organizational units	No. people
Operations	18
Administration	12
Planning and design	2
Client contact point	2
External	3

| Ratio of Manager to Subject-Matter Expert | | | |
Indicators	No.	Ratio 1	Ratio 2
Manager	5		
Subject-matter experts (SME)	29	17%	
Subtotal	34		
External	3		16%
Total	37		

TABLE 5.41 Network Management Process Indicators for Company 5 (Client Contact Point)

| | Client Contact Point Efficiency | | | | | |
| | Response delay to calls | | Response delay to voice mail | | Response delay to E-mail | |
Functions	Auto.	Man.	Auto.	Man.	Auto.	Man.
Handling troubles	22 secs.	60 secs.	5 mins.	30 mins.	1 hr.	5 hrs.
Handling changes	1 min.	5 mins.	15 mins.	45 mins.	2 hrs.	8 hrs.

TABLE 5.41 Network Management Process Indicators for Company 5 (Client Contact Point) (Continued)

Functions	Response delay to calls		Response delay to voice mail		Response delay to E-mail	
	Auto.	Man.	Auto.	Man.	Auto.	Man.
Handling orders	5 mins.	15 mins.	1 hr.	5 hrs.	2 hrs.	8 hrs.
Handling information enquiries	2 mins.	5 mins.	15 mins.	45 mins.	1 hr.	4 hrs.

Client Contact Point Grade of Service

Indicators	Outstanding	Good	Fair	Poor
Application problems		x		
Database problems		x		
Equipment problems		x		
Facility problems		x		
LAN problems		x		
WAN problems		x		
MAN problems			x	
Environmental problems			x	

TABLE 5.42 Network Management Process Indicators for Company 5 (Change Management)

Quality of Change Management

Time period	No. change requests	No. form complete	% complete	No. successful	% successful
Daily	30	28	93	28	93
Weekly	140	132	94	130	93
Monthly	560	548	98	552	99
Quarterly	1,600	1,546	97	1,562	98
Semiannually	3,300	3,250	98	3,270	99
Annually	6,800	6,770	99	6,695	98

Distribution Diagram of Change Duration

Duration	Severe priority 1 changes No.	%	Important priority 2 changes No.	%	Total	Percentage
15 min.	3	1.14	2	0.68	5	0.89
60 min.	6	2.28	6	2.14	12	2.14
2 hr.	20	7.58	24	7.86	44	3.00
4 hr.	28	10.60	30	10.36	58	10.36
8 hr.	72	27.27	102	31.07	174	31.07
2 days	40	15.15	48	15.71	88	15.71
4 days	45	17.05	47	16.43	92	16.43
7 days	35	13.26	30	11.61	65	11.61

Duration	Severe priority 1 changes No.	%	Important priority 2 changes No.	%	Total	Percentage
14 days	12	4.55	7	3.39	19	3.39
1 month	3	1.14	2	0.90	5	0.89
Total	264		296		560	100.00

Change Management Process Analysis

Indicators	Priorities No. priority 1	No. priority 2
Number of persons involved in changes	6	6
Number of steps in change management process	9	9
Number of automated steps	2	2
Number of manual steps	6	6
Number of semiautomated steps	1	1

TABLE 5.43 Fault Tracking and Monitoring Indicators for Company 5

Ratio of Proactive/Reactive Fault Detection

Time period	No. faults	No. proactive	No. reactive	% ratio proactive
Daily	70	10	60	14
Weekly	350	50	300	14
Monthly	1,400	400	1,000	28
Quarterly	5,200	1,100	4,100	21
Annually	21,000	4,500	16,500	21

Number of Troubles

Resources	Busy hours	Days	Weeks	Months	Quarter
Applications	8	30			
Batch		2			
Transactions		15			
Mail		13			
Equipment	5	20			
Hardware		8			
Software		12			
Facilities	5	20			
Primary circuits		17			
Backup circuits		3			
Backbone circuits		8			
Tail circuits		12			

TABLE 5.43 Fault Tracking and Monitoring Indicators for Company 5 (Continued)

Duration of Troubles

Resources	Total	15 mins.	1 hr.	4 hrs.	8 hrs.	1 week	4 weeks	More
Applications	30							
Batch	2			2				
Transactions	15		8	2	2	3		
Mail	13	1	6	5	1			
Equipment	20							
Hardware	8	1	2	5				
Software	12	1	3	4	3	1		
Facilities	20							
Primary circuits	17		1	3	10	4		
Backup circuits	3		1	1	1			
Backbone circuits	8		2	2	2	2		
Tail circuits	12		4	3	3	2		

Trouble Referrals

Date (time period)	No. of TT	No. filled in correctly	% ratio	No. referrals				
				1	2	3	4	more
August 4, 1994	70	68	97.14	65	2	2	1	
August 5, 1994	82	80	97.56	70	4	4	1	2
August 6, 1994	93	88	94.62	65	15	10	3	
August 7, 1994	65	60	92.30	58	6	1		
August 8, 1994	70	66	94.29	60	3	3	3	1
August 9, 1994	71	68	95.77	65	2	2	2	

Fault History Time Stamps

Indicators	Detected	Service restored	Vendor notified	Isolated	Repaired
Applications					
A	0:30	1:00	0:35	5:30	6:30
B	2:45	6:45	3:00	10:20	12:30
C	15:45	21:15	16:30	19:30	21:15
Equipment					
A	11:10	12:10	11:15	13:20	14:45
B	12:30	17:40	13:00	15:20	17:40
C	17:15	17:45	17:30	19:35	23:45
Facilities					
A	8:30	8:40	8:35	9:05	10:45
B	11:30	12:30	11:45	12:00	12:30
C	14:20	18:30	15:45	20:00	21:45

TABLE 5.44 Performance Monitoring Indicators for Company 5

	Availability				
Indicators	MTBF (hrs.)	MTTR (hrs.)	MTTI (hrs.)	MTTE (hrs.)	% avail.
Applications					
A	520	2	1	0.5	99.33
B	640	3	1	0.5	99.30
C	120	2	1	0.5	97.17
Equipment					
A	2,500	4	2	1	99.72
B	4,100	6	3	2	99.73
C	3,300	4	2	1	99.79
Facilities					
A	500	8	3	2	97.47
B	750	4	2	1	99.08
C	1,000	3	1	0.5	99.55

	Availability of Utilization Data				
				Instrument	
Indicators	Data avail.	Level of details	Consolidation periodicity	Manual	Monitor
Applications					
Batch	yes	day	day	yes	no
Online	yes	min.	hr.	no	yes
Mail	yes	hr.	day	yes	no
Equipment					
Server	no				
Clients	no				
Routers	no				
Multiplexer	yes	min.	hr.	no	yes
Facilities					
WAN	yes	hr.	day	no	yes
MAN	no				
LAN	yes	min.	hr.	no	yes

TABLE 5.45 Cost Indicators for Company 5

Cost items	Actual value (k)	Subtotal (k)	% of total
Hardware		20	19.05
WAN	5		
LAN	6		
Servers	6		
Clients	3		

TABLE 5.45 Cost Indicators for Company 5 (Continued)

Cost items	Actual value (k)	Subtotal (k)	% of total
Software		30	28.57
WAN	2		
LAN	3		
Servers	5		
Clients	5		
Application	15		
Infrastructure		20	19.05
Cabling	5		
Hubs	5		
Power supply	3		
Security surveillance	2		
Backup components	5		
Communications		25	23.81
Domestic lines	20		
International lines	5		
Value-added services			
Human resources		10	9.52
Management	3		
Operations	4		
Administration	2.5		
Planning and design	0.5		
Total		105	100.00

5.8 Summary

The valuable input examples in this chapter can be used for further analysis by the benchmarking team. The team and client used forms for network management functions, instruments, protocols, human resources, and skill levels. In addition, three different types of questionnaires were provided by the benchmarking team to the client. The final details for quantifying the performance of the client are supported by indicators defined by the benchmarking team.

Completing the forms and answering the questions could be arranged in different ways. To save on-premises time for the benchmarking team, information requests can be identified and sent out prior to starting the work. This tactic can work well for routine types of questions. Technical, operational, and strategic questions should be addressed in meetings.

The results of this chapter represent a tremendous amount of work. Chapter 6 and 7 use these entries for evaluating performance and comparing performance with other clients.

Data Consolidation and Reporting

6.1 Introduction

After the time-consuming phase of collecting data and information on processes, instruments, protocols, and people, the next phase begins, which covers compressing the data, interpreting information, and reporting results. The focus is still on the descriptive side of interpretation; the comparative site of the evaluation is addressed in Chapter 7.

On the basis of data analysis, both standard and special reports can be prepared and generated. Standard reports summarize questionnaires, processes, the use of instrumentation, protocols, and the definition of human responsibilities in the network management organization.

6.2 Standard Descriptive Reports

Standard descriptive reports include answers to the questions in the preliminary questionnaire and a summary of the answers to the on-site survey. Details are provided in the following subsections, which pull data from the examples provided in Chapter 5.

Preliminary questionnaire answers

This report contains the compressed answers following the structure of the eight items. The main purpose is to confirm and validate the answers on behalf of the clients. Using the entries in section 5.2, the following subsections provide a profile and attributes of the benchmarked company.

Company profile. The sample company is a financial institution. The networks support all communication forms, including data, voice, video, image, and electronic mail. The applications are very typical for a bank. The major locations span five continents. The branches are very similar to each other, thus all locations must be supported by all applications. The backbone of the network offers the T1-range of bandwidth; the access networks are slower, supporting bandwidth from 4.8 kbps up to 56 kbps.

Existing networks. The data networks are a combination of SNA, X.25, and frame relay. The architectures support SNA, FDDI, Ethernet, and token ring by the TCP/IP, SNA/SDLC, and IPX/SPX protocols. The leading operating systems include MVS, VM, OS/2, Unix, Windows, DOS, and Netware.

The voice networks represents a star structure. The video network uses the shared bandwidth from the voice and data networks. During peak video, the private voice network is supported by the public voice network. Backup strategies are manifold, including both physical and logical backups.

Transmission facilities for networks. The backbone network is exclusively digital and offers bandwidth in the T1 and E1 range. The access network is both digital and analog, depending on the geographical locations. Value-added services are part of the network; they provide packet switching, frame relay, and electronic mail from AT&T and Infonet.

Networking equipment. Principal devices in the wide area network are multiplexers, modems, packet switches, frame relay switches, and front-end processors. The local area networks are typically Ethernet or token ring connected locally by hubs and remotely by routers.

Personnel. A total of 160 persons manage the network. The biggest groups are operations and administration. This number is higher than expected considering the structure, number of managed objects, and architecture of the networks.

Costs. The distribution of costs between hardware, software, infrastructure, communications, and human resources does not show any surprises. Human resources consume approximately 40 percent of costs.

Network management. All network management functional groups, such as fault, configuration, performance, security, and accounting management, are supported. Instrumentation is average. Element management systems for WAN and LAN devices are in place. Integration is with NetView/390. The platform solution is just emerging with OpenView from Hewlett-Packard. Monitoring devices are available but not yet connected to the integration platform.

Outsourcing. The company is willing to outsource routine, day-to-day network management functions to a telecommunications provider. In this transition process, the company is willing to transfer up to 80 percent of its workforce to the outsourcer. The company, however, retains its rights for passively monitoring the network.

On-site survey answers

This report contains four segments, addressing the following areas:

- Part 1: Network management investments and organization (23 items)

- Part 2: Network management functions and problems (12 items)

- Part 3: Network management instrumentation (20 items)

- Part 4: Network management directions (14 items)

These reports contain the compressed answers to the questions asked in the questionnaire. Using the entries in section 5.3, the following results are presented.

Part 1: Network management investments and organization. The network management investments are medium. Economy of scale is gained with investments using payback analysis, present-value analysis, and return on investment. The information services manager maintains and supervises four separate groups responsible for networks, systems, databases, and applications management. The domestic and international networks are managed together by a staff of 160 persons. In building the network management team, team spirit is the highest priority. The company spends approximately 5 to 10 percent of its communication budget for network management. Lost revenues due to outages of WANs, MANs, and LANs are significant and can be used to build business cases for new investments.

Part 2: Network management functions and problems. This company fully agrees with the business areas described and explained in Chapter 3. The requests for modifications with functions and the allocation of functions to business areas are minimal.

The top three business areas are operations support, client control point, and fault monitoring. The most important challenges are considered to be technology changes, networking applications, new services, integrating instruments, real-time reporting, and the improvement of service quality in various countries.

Part 3: Network management instrumentation. The networks are centrally managed. The focal point for troubles and changes is the client control point. Network management services are offered on a 7 days per week, 24 hours per day basis. The responsibilities for the business areas are known at the highest level of the business area. Some conflicting allocations can exist at lower levels. Salary ranges are not unusual but are definitely not higher than industry average. Training in network management seems to be very well-organized, using various training sources. Still, the turnover of network management staff is high. Instrumentation is average; all principal instrumentation groups are represented. The experiences are best with element management systems.

If further integration is under consideration, network service providers are expected to play the role integrator. Outsourcers are judged by their financial strength, experience, and technology.

Part 4: Network management directions. Managing multiple vendors is very important for the company. Besides existing instruments, the company wants to use more OSI-based network management solutions within the next two to three years. NetView/390 and SNMP-based agents and managers are very important to the company. It has very clear expectations of what platform products should offer. These expectations include autodiscovery, SNMPv2 support, event management and correlation, automapping, SQL-capability, and distribution. The company is evaluating OpenView and NetView for AIX. One of the key evaluation criterion is the integration of device- and process-specific applications with the platform. The main purpose of the network is the validation of entries on behalf of the clients.

6.3 Consolidation and Reporting on the Basis of Forms

The forms completed during the on-site interview provide a means of evaluating and reporting on the company. Each of the forms, along with its accompanying report, is described in the following subsections.

Evaluation of the network management functionality matrix. This report contains 11 subreports that show the completed matrices for each function of the network management business areas:

- Client contact point
- Operations support
- Fault tracking
- Change control
- Planning and design
- Finance and billing
- Implementation and maintenance
- Fault monitoring
- Performance monitoring
- Security management
- Systems administration

The main purpose of this report is the validation of entries on behalf of the clients. The results are identical to those of sections 5.5 and 5.6. To improve the readability of the results, they are compressed here for Company B into Table 6.1. The table does not display the functions not supported.

TABLE 6.1 Company B Profile Using Form 3.1

	Fully supported	Partially supported	Not supported	Responsible party
Client Contact Point				
Receiving problem reports	x			HD
Handling calls	x			HD
Handling enquiries		x		HD
Opening and referring trouble tickets	x			HD, PC
Closing trouble tickets	x			HD, PC
Operations Support				
Determining problems using trouble tickets	x			NO, TS
Diagnosing problems	x			TS
Taking corrective actions	x			TS
Repairing and replacing	x			TS
Referring to third parties	x			TS
Backing up and reconfiguration		x		TS
Recovering		x		TS
Logging events and performing corrective actions		x		NO, TS

TABLE 6.1 Company B Profile Using Form 3.1 (Continued)

	Fully supported	Partially supported	Not supported	Responsible party
Fault Tracking				
Tracking manually reported or monitored faults	x			NO, TS, NA
Tracking the progress and escalation of problems if necessary		x		NO, TS, NA
Distributing information		x		AD
Referring problems	x			TS
Change Control				
Managing, processing, and tracking of service orders		x		CC
Routing service orders		x		CC
Supervising the handling of changes	x			CC
Planning and Design				
Analyzing needs		x		NA
Projecting application load		x		NA
Sizing resources				
Authorizing and tracking changes		x		CC
Finance and Billing				
Asset management		x		AD
Costing services	x			AD
Client billing		x		AD
Software license control		x		AD
Implementation and Maintenance				
Implementing change requests and work orders	x			TS
Maintaining resources		x		TS
Inspection		x		TS
Maintaining configuration database		x		TS, NA
Provisioning		x		TS
Fault Monitoring				
Monitoring system and network for proactive problem detection		x		NO, NA, TS
Opening additional trouble tickets	x			NO
Referring trouble tickets	x			TS, HD
Performance Monitoring				
Monitoring system and network performance		x		NA, TS
Optimizing, modeling, and tuning		x		NA, TS
Reporting on usage statistics and trends to management and users		x		NA, TS

	Fully supported	Partially supported	Not supported	Responsible party
Security Management				
Administration		x		SO
Detection	x			SO
Recovery	x			SO
Protecting the management systems		x		NA
Systems Administration				
Software version control				
Software distribution		x		AD, NA
Systems management		x		AD, NA
Management of names and addresses		x		AD
Applications management		x		AD

Key:

```
AD  =  Administrator
HD  =  Help Desk
NA  =  Network Analyst
NO  =  Network Operator
CC  =  Change Coordinator
PC  =  Problem Coordinator
TS  =  Technical Support
```

Detailed evaluation of network management functionality. On the basis of the previous report, the following subreports are generated:

- Frequency distribution for fully supported, partially supported, and unsupported functions for each of the 11 areas

- List of functions supported by the same organizational unit

- Percentages of functions where the responsibilities are split between organizational units

- Percentages of functions supported by the same group

Table 6.2 shows the frequency distribution. In general, fault-related activities are well-supported. Weaknesses are very obvious in the areas of planning and design, finance and billing, performance management, and systems administration.

Table 6.3 lists the functions supported by the same organizational unit. It is obvious that one organizational unit is in charge of multiple functions, which is usual in this industry. On the other hand, there are areas where responsibilities are split between different organizational units. Three organizational units are in charge of the network management functions.

TABLE 6.2 Frequency Distribution for Network Management Functions

Business areas	Level of support			Total
	Fully supported	Partially supported	Not supported	
Client contact point	4	1	3	8
Operations support	5	3		8
Fault tracking	2	2		4
Change control	1	2		3
Planning and design		3	5	8
Finance and billing	1	3	3	7
Implementation and maintenance	1	4		5
Fault monitoring	2	1		3
Performance monitoring	2	1	2	5
Security management	2	2	1	5
Systems administration		4	3	7
Total	20	26	17	63

The functions are:

- Tracking manually reported or monitored faults
- Tracking the progress and then escalating problems when necessary
- Monitoring the system and network for proactive problem detection

which corresponds to 4.75 percent of the total number of network management functions (63).

TABLE 6.3 List of Functions Supported by Same Organizational Unit

Organizational unit	Functions
Network manager	Overall responsibility
Control coordinator	Managing, processing, and tracking service orders Routing service orders Supervising handling of changes Authorizing and tracking changes
Problem coordinator	Opening and referring trouble tickets Closing trouble tickets
Help desk	Receiving problem reports Handling calls Handling enquiries Opening and referring trouble tickets Closing trouble tickets Referring trouble tickets

Organizational unit	Functions
Network operator	Determining problems using trouble tickets Logging events and performing corrective actions Tracking manually reported or monitored faults Tracking progress and escalation of problems Conducting proactive monitoring Opening additional trouble tickets
Network analyst	Tracking manually reported or monitored faults Tracking progress and escalating problems Performing needs analysis Projecting application load Maintaining configuration database Conducting proactive monitoring Monitoring system and network performance Optimization, modeling, and tuning Reporting Protecting the management system Distributing software Systems management
Technical support	Determining problems Diagnosing problems Taking corrective actions Performing repair and replacement Referring to third parties Performing backup, reconfiguration, and recovery Logging events and corrective actions Tracking manually reported or monitored faults Tracking progress and escalating problems Implementing change requests and work orders Maintaining resources Conducting inspection Maintaining configuration database Provisioning Monitoring performance Optimization, modeling, and tuning Reporting
Security officer	Administering security violations Detecting security violations Performing recovery
Administrator	Distributing information Managing assets Performing costing services Billing clients Distributing software Managing system Managing names and addresses Managing applications

Two organizational units are in charge of the following network management functions:

- Opening and referring trouble tickets
- Closing trouble tickets
- Determining problems by handling trouble tickets
- Logging events and corrective actions
- Maintaining configuration database
- Referring trouble tickets
- Monitoring system and network performance
- Optimizing, modeling, and tuning
- Reporting on usage statistics and trends
- Distributing software
- Managing the systems

which corresponds to 17.45 percent of the total number of network management functions (63). The responsibility question is not yet clear in slightly more than 22 percent of network management functions, which can cause serious delays in executing functions. In particular, change and fault management functions can be heavily impacted. It is recommended to migrate to a single responsibility structure.

Overview of network management instrumentation. This report lists the instruments currently used or planned. The main purpose of this report is the validation of entries on behalf of the clients. Table 6.4 shows the results of using network management instruments. This list is a compressed version of Form 3.2 used in section 5.6. Only the supported instruments and their status (in use or planned) are shown.

TABLE 6.4 Network Management Instrumentation Overview (Form 3.2)

Instruments	In use	Planned	Owner	Support of functional areas
Integrators				
Manager of managers	x		NA, NM, NO	CCP, OS, FM, FT
Management platforms		x	NA, NM, NO	CCP, OS, FM, FT
Element management systems for WANs				
Modems	x		TS, NO	CCP, OS, FM, FT
Multiplexers	x		TS, NO	CCP, OS, FM, FT
Matrix switches	x		TS, NO	CCP, OS, FM, FT

Instruments	In use	Planned	Owner	Support of functional areas
Element management systems for LANs				
Bridges	x		TS, NO	CCP, OS, FM, FT
Routers	x		TS, NO	CCP, OS, FM, FT
Hubs	x		TS, NO	CCP, OS, FM, FT
Monitors and analyzers				
WAN monitor	x		TS, NA	FM, PM
LAN monitor	x		TS, NA	FM, PM
WAN analyzer	x		TS, NO	FM, PM
LAN analyzer	x		TS, NO	FM, PM
Software monitor	x		NA	FM, PM
Security management systems				
Protection of systems and networks	x		SO	SM
Protection of management systems		x	NA	SM
Administration instruments				
Documentation systems	x		AD, NM	IM, SA, PD, FT, CCP
Modeling instruments		x	NA	PD, PM
Presentation tools		x	NA	FT, FM, PM
Report generators	x		NA	PM, FB, PD
Troubletracking tools	x		AD, NO	OS, CCP, FT
Software distribution tools		x	AD	SA
Software licensing tools		x	AD	FB
Database tools				
Databases	x		NA	SA
Client contact point instruments				
Pager	x		TS	CCP, OS, FM

Key:

AD	=	Administrator
CCP	=	Client Contact Point
FB	=	Finance and Billing
FM	=	Fault Monitoring
FT	=	Fault Tracking
IM	=	Implementation and Maintenance
NA	=	Network Analyst
NM	=	Network Manager
NO	=	Network Operator
OS	=	Operations Support
PD	=	Planning and Design
PM	=	Performance Management
SA	=	Systems Administration
SM	=	Security Management
SO	=	Security Officer
TS	=	Technical Support

Overview of network management protocols. This report lists the protocols currently used or planned. The main purpose of this report is the validation of entries on behalf of the clients. Table 6.5 shows the results of using network management protocols. This list is a compressed version of Form 3.3 used in section 5.6. Only the supported protocols and their status (in use or planned) are shown.

Human resources that support network management processes and instruments. This report identifies established and planned network management job titles. It also identifies other job titles within the network management organization when they differ from the list provided by the benchmarking company. The main purpose of this report is the validation of entries on behalf of the clients. Table 6.6 shows the results of which job titles have been established or are being established within the network management organization. This list is a compressed list of Form 3.4 used in section 5.6. Only the established or planned job titles with their special names are shown.

TABLE 6.5 Network Management Protocol Overview

Network Management protocols	Established	Planned	Other protocols
Proprietary protocols			
NMVT	x		
MSU		x	
Novell	x		
SNMPv1	x		
SNMPv2		x	
RMON		x	

TABLE 6.6 Human Resources Supporting Network Management (Form 3.4)

Job titles	Established	Planned	Other job titles
Network manager	x		NM
Inventory and assets coordinator		x	AD
Cable management administrator		x	AD
Change coordinator	x		CC
Problem coordinator	x		PC
Order processing and provisioning coordinator		x	AD
Network operations control supervisor	x		NM
Client contact point operator	x		HD
Network operator	x		NO
Network technician	x		TS
Network performance analyst	x		NA
Database administrator	x		AD
Modeling coordinator		x	NA
Security management supervisor		x	SO
Security officer	x		SO

Job titles	Established	Planned	Other job titles
Costing specialist	x		AD
Accounting clerk		x	AD
Service-level coordinator		x	AD
Business planner		x	NM
Technology analyst	x		NA

Key:

AD = Administrator
CC = Change Coordinator
HD = Help Desk
NA = Network Analyst
NM = Network Manager
NO = Network Operator
PC = Problem Coordinator
SO = Security Officer
TS = Technical Support

Compressed summary of activity logs. This report contains entries for all the network management functions being analyzed in the client's environment. Usually, the basic operational analysis includes the activity logs for high-priority functions only. The logs are separated by shifts. The main purpose of this report is the validation of entries on behalf of the clients.

Activity logs are not typically reported in a special format, but simply sorted by resources, geographical areas, users, or time. In particular, when the logical sequence of certain events is important, logs could become very useful. Figure 6.1 shows a simple example for troubleshooting a global problem for a multinational corporation based in the United States.

Matrix of network management functions and personnel. This report lists the allocation of management functions grouped around network management business areas to human resources that support network management. The present status is characterized in Tables 6.7 and 6.8. Table 6.7 gives an overview of fully supported functions, including those of the supporting group. Table 6.8 offers the same for partially supported functions.

Matrix of network management functions and instruments. This report lists the allocation between management functions grouped around network management business areas and network management instruments. The correct use of instruments is extremely important. Based on Form 3.2 used in section 5.6, Table 6.9 shows the allocation of instruments to functional areas. This allocation matrix can be for two different purposes:

- Identification of the right instruments for each supported network management function.

- Identification of network management functions where the same instrument can be used.

Date: March 22, 1994

05:07   HP Openview indicates alarms form the Spain side of transmission connections in the Network Management Center of service provider.
Trouble ticket opened.
Attempt to restore service by using dial backup facility instead of failed circuit.

05:10   Attempt fails, Spain does not respond.
Service provider notifies customer.

05:12   Joint conclusion: because primary and secondary services failed simultaneously, problem expected to be on customer premise in Spain.

05:32   Attempt to contact customer in Spain without success.
European management center has been alarmed by the U.S.

05:44   European management center tests multiplexer in Spain; it is not responding.
Problem is being escalated to British Telecom and Telefonica.

06:12   British Telecom confirms no problem on British side.
Customer receives status from service provider.

07:34   Telefonica escalates problem to multiplexer vendor in Spain.
Simultaneously, U.S service provider dispatches technician to premise of customer.

09:46   Technician from service provider arrives and starts testing on-premise circuits.

09.55   Vendor of multiplexer arrives on site and starts testing equipment.

11:15   Vendor diagnoses problem; both power supply units failed, along with protection fuses.
Customer receives status in both European and U.S. locations.

13:25   Spain node is restored and primary link is operational.
Trouble ticket for primary problem is closed.
Vendor requested by service provider to replace multiplexer in Spain.

14:00   Service provider starts new dispatch to find problem with dial backup.
Engineers assume problem caused by looped condition at router at U.S. side.
Trouble ticket is opened and referred to vendor of router.

Figure 6.1 Compressed summary of activity logs

TABLE 6.7 Fully Supported Network Management Functions and Personnel

	Fully supported	Partially supported	Not supported	Responsible party
Client Contact Point				
Receiving problem reports	x			HD
Handling calls	x			HD

	Fully supported	Partially supported	Not supported	Responsible party
Opening and referring trouble tickets	x			HD, PC
Closing trouble tickets	x			HD, PC

Operations Support

	Fully supported	Partially supported	Not supported	Responsible party
Determining problem using trouble tickets	x			NO, TS
Diagnosing problems	x			TS
Taking corrective actions	x			TS
Repairing and replacing	x			TS
Referencing to third-parties	x			TS

Fault Tracking

	Fully supported	Partially supported	Not supported	Responsible party
Tracking manually reported or monitored faults	x			NO, TS, NA
Referring problems	x			TS

Change Control

	Fully supported	Partially supported	Not supported	Responsible party
Supervising handling of changes	x			CC

Finance and Billing

	Fully supported	Partially supported	Not supported	Responsible party
Costing services	x			AD

Implementation and Maintenance

	Fully supported	Partially supported	Not supported	Responsible party
Implementing change requests and workorders	x			TS

Fault Monitoring

	Fully supported	Partially supported	Not supported	Responsible party
Opening additional trouble tickets	x			NO
Referring trouble tickets	x			TS, HD

Performance Monitoring

	Fully supported	Partially supported	Not supported	Responsible party
Monitoring system and network performance	x			NA, TS
Reporting on usage statistics and trends	x			NA, TS

Security Management

	Fully supported	Partially supported	Not supported	Responsible party
Detection	x			SO
Recovery	x			SO

Key:

HD = Help Desk
NA = Network Analyst
NO = Network Operator
SO = Security Officer
TS = Technical Support

TABLE 6.8 Partially Supported Network Management Functions and Personnel

	Fully supported	Partially supported	Not supported	Responsible party
Client Contact Point				
Handling enquiries		x		HD
Operations Support				
Backing up and reconfiguration		x		TS
Recovering		x		TS
Logging events and corrective actions		x		NO, TS
Fault Tracking				
Tracking the progress and escalation of problems if necessary		x		NO, TS, NA
Distributing information		x		AD
Change Control				
Managing, processing, and tracking service orders		x		CC
Routing service orders		x		CC
Planning and Design				
Analyzing needs		x		NA
Projecting application load		x		NA
Authorizing and tracking changes		x		CC
Finance and Billing				
Asset management		x		AD
Client billing		x		AD
Software license control		x		AD
Implementation and Maintenance				
Maintaining resources		x		TS
Inspection		x		TS
Maintaining the configuration database		x		TS, NA
Provisioning		x		TS
Fault Monitoring				
Monitoring the system and network for proactive problem detection		x		NO, NA, TS
Performance Monitoring				
Optimizing, modeling, and tuning		x		NA, TS

	Fully supported	Partially supported	Not supported	Responsible party
Security Management				
Administration		x		SO
Protecting the management systems		x		NA
Systems Administration				
Software distribution		x		AD, NA
Systems management		x		AD, NA
Management of names and addresses		x		AD
Applications management		x		AD

Key:

AD = Administrator
CC = Change Coordinator
HD = Help Desk
NA = Network Analyst
NO = Network Operator
SO = Security Officer
TS = Technical Support

TABLE 6.9 Network Management Functions and Instruments

Instruments	Functions										
	CCP	OS	CC	FM	FT	PM	SM	IM	SA	FB	PD
Integrators											
Manager of managers	x	x		x	x						
Management platforms	x	x		x	x						
EMS-WAN											
Modems	x	x		x	x						
Multiplexer	x	x		x	x						
Matrix switches	x	x		x	x						
EMS LAN											
Bridges	x	x		x	x						
Routers	x	x		x	x						
Hubs	x	x		x	x						
Monitors and analyzers											
WAN monitor				x	x						
LAN monitor				x	x						
WAN analyzer				x	x						
LAN analyzer				x	x						
Software monitor				x	x						
Security management systems											
Tools for systems and networks								x			
Tools for management systems								x			

TABLE 6.9 Network Management Functions and Instruments (Continued)

Instruments	CCP	OS	CC	FM	FT	PM	SM	IM	SA	FB	PD
Administration											
Documentation	x			x				x	x		x
Modeling					x						x
Presentation			x	x	x						
Reports					x					x	x
Trouble tracking	x	x		x							
SW distribution									x		
SW licensing									x		
Database Tools											
Databases			x						x		
Client contact point											
Pager	x	x			x						

Key:
CC   = Change Control
CCP = Client Contact Point
FB   = Finance and Billing
FM   = Fault Monitoring
FT   = Fault Tracking
IM   = Implementation and Maintenance
OS   = Operations Support
PD   = Planning and Design
PM   = Performance Management
SA   = Systems Administration
SM   = Security Management

Matrix of network management personnel and instruments. This report lists the allocation of network management instruments to human resources supporting network management. Based on Form 3.2 used in section 5.6, Table 6.10 summarizes the responsibilities of human resources for network management instruments. This table can be changed into a matrix to significantly improve readability.

The final example, shown in Table 6.11 on page 218, is the responsibility/skill matrix for selected well-supported network management areas. Based on Form 3.5 used in section 5.6, the following business areas are revisited:

- Operations support
- Fault tracking
- Fault monitoring

As can be seen, the skills available are sufficient to support the individual network management functions. Chapter 7 revisits these tables again and compares them to targeted skill sets.

TABLE 6.10 Network Management Personnel and Instruments

Human resources	Network management instruments
Network manager	Manager of managers Management platforms Documentation tools
Network operator	Manager of managers Management platforms Modem manager Multiplexer manager Matrix switch manager Bridge manager Router manager Hub manager WAN analyzer LAN analyzer Trouble-tracking tool
Network analyst	Manager of managers Management platforms WAN monitor LAN monitor Tool for protecting management system Modeling instrument Presentation tool Report generator Databases
Technical support	Modem manager Multiplexer manager Matrix switch manager Bridge manager Router manager Hub manager WAN monitor LAN monitor WAN analyzer LAN analyzer Pager
Administrator	Documentation system Trouble-tracking tool Software distribution tool Software licensing tool
Security officer	Protection tool for systems and networks
Help desk	Trouble-tracking tool Pager

TABLE 6.11 Selected Examples for Responsibility/Skill Matrix

Functions	Functional knowledge	General telecom	In-depth telecom	Instrument knowledge	Personal communication	Project management
Operations Support						
Determining problems by handling trouble tickets	x		x	x		
Diagnosing problems	x		x	x	x	
Taking corrective action	x		x	x		
Repairing and replacing	x		x	x		
Referring to third parties	x		x	x		
Backing up and reconfiguring	x		x	x	x	
Recovering	x		x	x		
Logging events and corrective action	x	x		x		
Fault Tracking						
Tracking manually reported or monitored faults	x	x				
Tracking progress and escalating problems if necessary	x	x		x		
Distributing information	x					
Referring problems	x		x		x	
Fault Monitoring						
Monitoring system and network for proactive problem detection	x		x	x		
Opening additional trouble tickets	x	x		x		
Referring trouble tickets	x	x		x		

6.4 Specific Descriptive Reports

The specific reports contain answers to the detailed survey questions grouped around specific functions of network management business areas. The principal emphasis is on the network management processes and indicators.

The survey questions of section 4.2 (Appendix C) address each individual function of each business area. Depending on the priorities of the corporation, certain functions are investigated in greater depth.

In the case of provisioning, change management, fault management, performance tuning, security management, and accounting, the chains and interrelationship of functions are evaluated in depth. Individual customer solutions are compared against sample process flows, representing the most likely solutions.

The second group of reports discusses the actual indicators allocated to individual management areas. The majority of these indicators offer quantitative values that are collected, consolidated, correlated, and processed during benchmarking. Examples are shown in section 6.5.

6.4.1 Network management functions

Specific reports about network management functions by business areas include the following:

- Client contact point
- Operations support
- Fault tracking
- Change control
- Planning and design
- Finance and billing
- Implementation and maintenance
- Fault monitoring
- Performance monitoring
- Security management
- Systems administration

Client contact point. This report provides a detailed summary of results of the in-depth questionnaire about the client contact point. Subreports address the following:

- Receiving problem reports
- Handling calls
- Handling enquiries

- Receiving change requests
- Handling orders
- Making service requests
- Opening and referring trouble tickets
- Closing trouble tickets

Operations support. This report provides a detailed summary of results of the in-depth questionnaire about operations support related questions, including second- and third-level problem resolution. Subreports address the following:

- Determining problems using trouble tickets
- Diagnosing problems
- Taking corrective actions
- Repairing and replacing
- Referencing to third parties
- Backing up and reconfiguration
- Recovering
- Logging events and performing corrective actions

Fault tracking. This report provides a detailed summary of results of the in-depth questionnaire about fault tracking related questions. Subreports address the following:

- Tracking manually reported or monitored faults
- Tracking progress and escalation of problems if necessary
- Distributing information
- Referring problems

Change control. This report provides a detailed summary of results of the in-depth questionnaire about change control related questions. Subreports address the following:

- Managing, processing, and tracking of service orders
- Routing service orders
- Supervising the handling of changes

Planning and design. This report provides a detailed summary of results of the in-depth questionnaire about planning and design related questions. Subreports address the following:

- Analyzing needs
- Projecting application load
- Sizing resources
- Authorizing and tracking changes
- Raising purchase orders
- Producing implementation plans
- Establishing company standards
- Maintaining quality assurance

Finance and billing. This report provides a detailed summary of results of the in-depth questionnaire about finance and billing related questions. Subreports address the following:

- Asset management
- Costing services
- Client billing
- Usage and outage collection
- Calculation of rebates to clients
- Bill verification
- Software license control

Implementation and maintenance. This report provides a detailed summary of results of the in-depth questionnaire about implementation and maintenance related questions. Subreports address the following:

- Implementing change requests and work orders
- Maintaining resources
- Inspection
- Maintaining the configuration database
- Provisioning

Fault monitoring. This report provides a detailed summary of results of the in-depth questionnaire about fault monitoring related questions. Subreports address the following:

- Monitoring system and network for proactive problem detection
- Opening additional trouble tickets
- Referring trouble tickets

Performance monitoring. This report provides a detailed summary of results of the in-depth questionnaire about performance monitoring related questions. Subreports address the following:

- Monitoring system and network performance
- Monitoring service-level agreements
- Monitoring third-party and vendor performance
- Optimizing, modeling, and tuning
- Reporting on usage statistics and trends to management and users

Security management. This report gives a detailed summary of results of the in-depth questionnaire about security management related questions. Subreports address the following:

- Threat analysis
- Administration
- Detection
- Recovery
- Protecting the management systems

Systems administration. This report provides a detailed summary of results of the in-depth questionnaire about systems administration related questions. Subreports address the following:

- Software version control
- Software distribution
- Systems management
- Administration of user-definable tables
- Local and remote configuration of resources
- Management of names and addresses
- Applications management

6.4.2 In-depth analysis of network management functions

It is very rare that benchmarks go into detail for all 63 functions defined for the 11 business areas of network management. Typically, just the high-priority functions are investigated in greater depth. Based on the entries of section 5.4, 19 functions (approximately 30 percent) were audited for the final details. The compressed versions could be incorporated into the final presentation without major changes, but some editing. It is very important to include conclusions and recommendations for each of the functions. Examples are listed in the following subsections.

Receiving problem reports. Many problems are reported manually to the client contact point. No well-established documentation form exists for registration and tracking. Receiving reports is supported 7 days per week, 24 hours per day, but the level of automation is very low. Subjectivity dominates in asking and interpreting symptoms.

Handling calls. The majority of problems are reported over the phone rather than electronically. The Help Desk handles calls 7 days per week, 24 hours per day. No well-established documentation form exists for registration and tracking with the exception of station message detailed recordings (SMDRs). The level of automation is very low. Interactive voice response has not yet been implemented. Subjectivity dominates in asking and interpreting symptoms.

Receiving change requests. Requests are submitted with change request forms. Manual review exists for impacts, completeness, and scheduling alternatives. A frequent dialogue exists between the requester and the operator receiving the requests. No special support instruments are in use. The documentation is supported by forms very similar to trouble tickets. The lack of quality of incoming information is due to the incompleteness of forms.

Opening and referring trouble tickets. Tickets are opened for all types of problems. Tickets are referred after five minutes by the Help Desk to other areas. This service is supported 7 days per week, 24 hours per day. PNMS III from Peregrine supports the documentation and tracking of trouble tickets. All information is entered manually; no automation in copying attributes from other databases or files exists. The quality of incoming information is low due to inaccuracies in reporting troubles by users. Referrals require intelligent decisions by the Help Desk.

Closing trouble tickets. Tickets are closed after confirmation of functionality by the user who reported the problem. This service is supported 7 days per week, 24 hours per day. PNMS III from Peregrine supports this activity. The closing procedure is still manual; all relevant information from the resolution process, such as test results and measurements, are attached to the ticket.

Determining problems using trouble tickets. Problem determination is supported for facilities, equipment, databases, and applications. This function identifies the problem and all components impacted. Time limitations have been established on the basis of estimated impacts of managed objects. This function is supported 7 days per week, 24 hours per day. A number of instruments exist, such as monitors, element management systems, and the trouble-ticketing product that supports problem determination. Monitoring and alarm management are automated. Alarm correlation is manual. To facilitate decision-making, escalation procedures are available. The quality of input is usually very good.

Diagnosing problems. This function determines why the components experience problems. Scope includes all managed objects with emphasis on facilities and equipment. Both second- and third-level support is offered with this function. The basis of problem diagnosis is trouble tickets referred by operators and expanded by notes, measurement results, and diagnostics. PNMS III is the support instrument for documentation. Monitors and analyzers are helpful in diagnosing and eliminating problems. This activity is highly sophisticated, and experience is absolutely necessary.

Tracking monitored faults. All faults with open trouble tickets are tracked. The lifecycle of trouble tickets depends on service-level agreements. Depending on the managed objects, time limits for resolution have been introduced. Status review is periodic, but high-priority troubles overtake periodic status evaluation. Documentation instruments include trouble tickets (PNMS III), fax, and E-mail. The automation level is low, but simple statistics are reported automatically. There are multiple priority levels depending on managed objects. The quality of incoming information depends on the completeness of the trouble tickets.

Supervising the handling of changes. This function concentrates on changes for facilities, equipment, databases, and applications. It incorporates fixes, immediate changes, and periodic changes. Change-request forms are the basis for documentation. The product is PNMS III, which allows basic automation of sorting, processing, and reporting of trouble tickets. Execution priorities are higher for fixes and immediate changes. The quality of incoming information depends on the completeness of change requests. The forms are very well prepared by the product.

Authorizing and tracking changes. Requests are authorized for facilities, equipment, databases, and applications. Expedited changes, which include fixes and immediate changes, are handled within one working day. PNMS III is the basis of the documentation and exchange of information between business areas. The automation level is very low. All change requests arrive from a single point of contact.

Establishing company standards. Standards include network management and guidelines for purchasing hardware, software, and applications. This function offers two reviews per year. Paper-based reports are used for documentation. There is no automation. Input is from various standards bodies, including the Network Management Forum, OSF, DMTF, the FDDI Forum, and the ATM Forum. Information from the Internet Advisory Board is also considered, which is very important for SNMP. The results of this function are used for negotiations with product vendors.

Costing services. All equipment and facilities are included in the accounting structure. Documentation is computer-based, using word processors and spreadsheets. Products in use are SMF (IBM), SAS for statistical evaluations, and Excel spreadsheets for ad hoc evaluation and reporting. Automation concentrates on processing mass data and distributing information to other groups.

Sampling software licenses. This function includes all operating systems, databases, and applications from all vendors. All documentation and processing are in Excel spreadsheets. No automation exists yet with this function. Licenses are sampled for violations. The source of information includes inventory and utilization data. The input quality is low due to incomplete and inaccurate data.

Implementing change requests and work orders. The scope of this function includes facilities, equipment, databases, and applications. The basis of documentation is partially paper-based and partially electronic workorders. The product in use is a "home-grown" extension of PNMS III. Distribution of workorders is still manual. The quality of incoming information is very good.

Monitoring system and network for proactive problem detection. Proactive monitoring concentrates on facilities, equipment, and systems. Once measured, evaluation and interpretation of results is accomplished in real time. Both eventing and polling techniques are used for information collection. Data collection is highly automated. A large number of instruments exist that are used for collection, interpretation, and visualization. The most important ones are NetView/390, TimeView, NetView AIX, SunNet Manager, GMF (IBM), and various WAN and LAN monitors and analyzers. Priorities are high for backbone facilities and equipment that interconnects various LANs. The accuracy of incoming information is sufficient. Correlating alarms is still manual.

Monitoring service-level agreements. The function concentrates on the full content of service-level agreements. In particular, service indicators are tracked and compared with expected values. MTBF, MTTR, and utilization indicators are measured on monthly basis. Written reports are the basis of evaluation. Monitors are used for measurements, SAS for gathering statistics, and Excel for ad hoc reports. Compressing data is automatic, but the interpretation is still manual. Escalation procedures are preprogrammed and followed by executing the function. Expedited escalation is required when the same indicator has been violated three times in sequence.

Optimizing, modeling, and tuning. Target indicators with this function are availability, response time, workload, and resource utilization. This function is mission-driven, with each execution taking three to four weeks on average. Project files and optimization handbooks support the documentation. A variety of instruments support this function, including monitors, element management systems, databases, reporting software, and modeling packages for WANs and LANs. This activity is highly creative, offering practically no opportunities for automation. The quality of incoming information is very different; workload estimates for modeling are very inaccurate, but monitored indicators are very accurate.

Protecting the management systems. This function concentrates on the protection of network management products by authorization and authentication of use. This protection is continuous and supported by partitioning access rights of network management personnel. Logs and written reports are used as documentation tools. Special products are in use, but Unix-based features are used for authorization and personalized data for authentication. Detection is automatic, but actions are not. Decision-making criteria are based on the severity of impacts. Configuration isolations are considered the most severe. The quality of input data is satisfactory. NetView/390 is protected by resource access control facility (RACF) in the mainframe.

Software distribution. This function is responsible for the distribution of operating systems and applications. Presently, the distribution is supported by the night shift only. Because of the large number of servers and end-user devices, the "push" technology is used. As instruments, IBM's distribution tools and a project scheduling package are in use. The automation level is high during distribution but low during the preparation of packages in the centrally located depot.

6.5 Specific Reports about Network Management Processes

These reports are generated only when the company under benchmark operates using similar processes to those described in Chapter 3. During the data collection phase, the conformance toward these models is investigated in detail. The conformance reports include the following:

- The preprovisioned change management process
- How the client-reported faults process is handled

- How the network/system reported fault process is handled
- The performance tuning process
- Security control process
- Non-preprovisioned change management process
- The costing and billing process

Chapter 3 gives an in-depth treatment of principal network management processes using various flowcharts. Certain divisions and groups of the sample company do have operational guidelines, but these are less sophisticated than the recommended process flows in Chapter 3.

There are three categories of conformance. If processes for these targeted areas are not in place, conformance reports are not generated.

- High-level conformance is applied if the functions and flows are more than 90 percent identical.
- Middle-level conformance is applied if the functions and flows are more than 50 percent similar.
- Low-level conformance is applied if just a few functions or flows are similar.

Chapter 5 presents two examples for processes: addressing change and fault management. The evaluation and comparison with the expected flowcharts show the following results.

Change management process

- Flowchart is available (Figure 5.1)
- The flowchart addresses the major steps of the change management process; the minor steps are omitted.
- There is only one initiator: the change requester.
- The flowchart is limited to the functions of the change coordinator; other groups are not identified or involved.
- The information exchange between responsible areas is neither identified nor explained.
- The level of detail is at least two steps higher than the expected level by the process model.

The conclusion of the benchmarking team is **low-level conformance**.

Client-reported fault management process

- Flowchart is available (Figure 5.2)
- The flowchart addresses the major steps of the fault management process; minor steps are omitted.

- The initiators of the process are clearly defined.

- The responsible groups are identified, but the allocation of tasks is not clearly defined.

- The information exchange between responsible areas is neither identified nor explained.

- The level of detail is one step higher than the expected level by the process model.

The conclusion of the benchmarking team is **middle-level conformance**.

6.6 Specific Network Management Indicator Reports

On the basis of trouble tickets, monitored data, logs, and special manual records, indicators can be computed and reported. Most indicators reported here can also be used in the next chapter as comparative indicators. The interpretation of some indicators requires prior experience and special network management skills.

The following indicator groups are reported.

Generic indicators. This group contains indicators of service quality, user satisfaction, scalability of network management services, and maintainability of service indicators.

Figure 6.2 shows the quality of attended operations supported by a distribution function. The form of the curve is acceptable; the *good* entry dominates, *outstanding*, *fair*, and *poor* keep balance.

Organizational indicators. Most reports of this group deal with staffing using details about staff size, the ratio between managers and subject-matter experts, educational level, skill level, and qualifying experience.

The distribution of human resources between various network management functional areas is shown in Figure 6.3. At first glance, the distribution seems to be well balanced, but the number of administrative persons might be slightly higher than expected.

The ratio between managers and subject-matter experts is displayed in Figure 6.4. Excluding external experts, this ratio is 13 percent, which is in the expected range. Including external work force, the ratio drops to 11 percent, which is relatively low but still in range (10 percent is the low indicator).

Network management processes. In addition to the process-conformance reports, specific process indicators help quantify process quality. Figure 6.5 shows the grade of service at the client contact point, supported by a distribution function. The *good* response dominates, but the *poor* entry is slightly higher than expected. *Outstanding* and *fair* balance.

Frequency

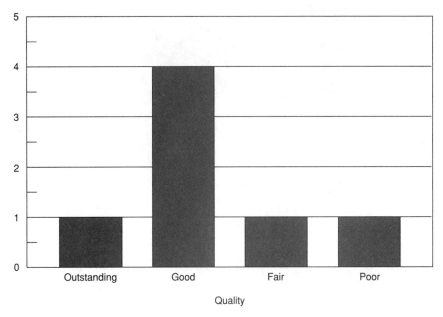

Figure 6.2 Quality of attended operations

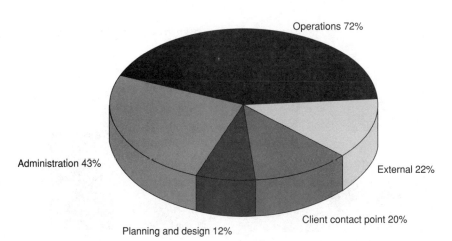

Figure 6.3 Distribution of human resources

The results of change management quality investigations are very favorable. Both curves shown in Figure 6.6 for the completion ratio and the successful ratio are satisfactorily high and show great stability at longer observation intervals.

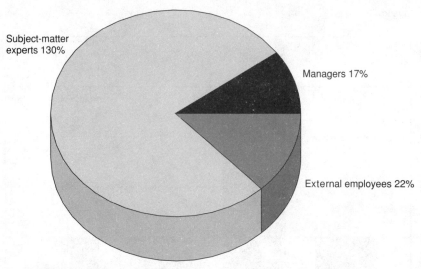

Figure 6.4 Ratio of manager, subject-matter experts, and external consultants

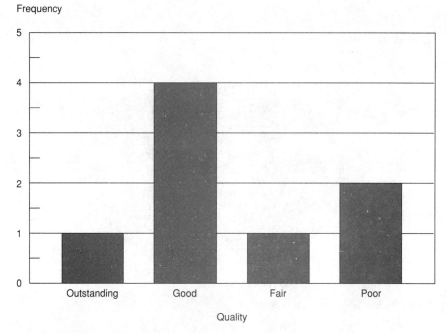

Figure 6.5 Grade of service at client contact point

Figures 6.7 and 6.8 show the distribution diagrams for the duration of priority 1 and priority 2 changes. The form of the distribution functions are very similar; the usual Gaussian distribution is clearly recognizable.

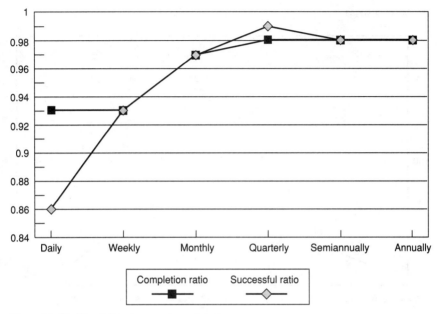

Figure 6.6   Quality of change management

Number of changes

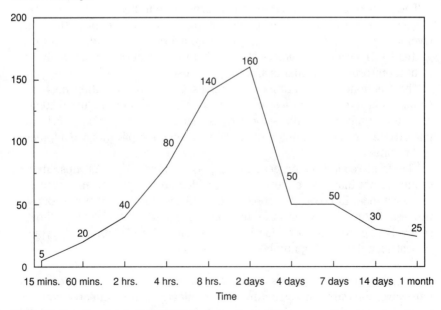

Figure 6.7   Duration distribution of Priority 1 changes

Number of changes

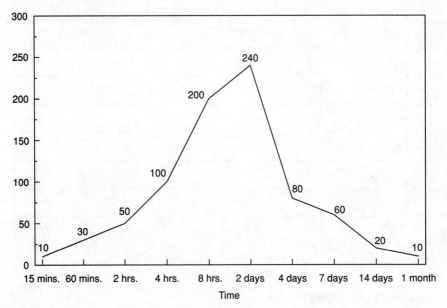

Figure 6.8 Duration distribution of Priority 2 changes

The proactive fault detection ratio is presented in Figure 6.9. A peak exists on the monthly base. The benchmarking team cannot explain this peak. Over longer observation periods, such as quarters and beyond, the ratio migrates to 21 percent, which is still far from expectations. Most likely, the right monitoring instruments are not yet in use.

The frequency and duration of troubles is vital to operating networks. Outages cause significant losses to the larger organization. Figure 6.10 analyzes the number of troubles for various managed objects. Figure 6.11 displays three typical functions about fault duration for applications, equipment, and facilities.

The fault-resolution process is supported by multiple groups. If one group cannot diagnose or resolve the problem, other groups are contacted for assistance. Referrals of problems extend the duration of the resolution process. Figure 6.12 shows the distribution of *1, 2, 3, 4,* and *More than 4* referrals of problems for a sample of six entries. The distribution is typical, indicating a dominating number for *1* referral.

Cost.   A separate category is established for costs and is broken down into hardware, software, infrastructure, communications, and human resources. In most companies, these indicators take top priority.

% Proactive fault detection

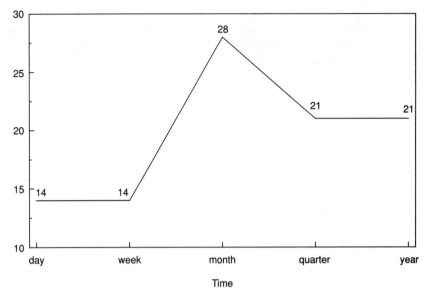

Figure 6.9 Proactive fault detection ratio over time

Number of troubles

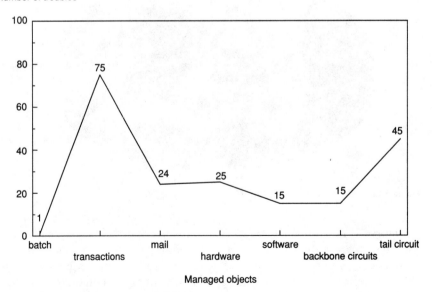

Figure 6.10 Number of troubles by managed objects

Percent of distribution
of fault duration

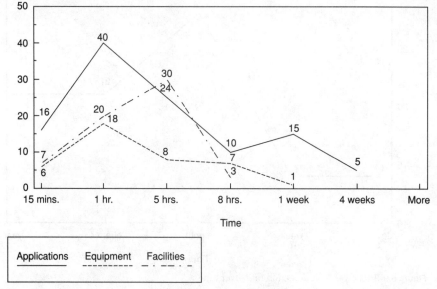

Figure 6.11 Distribution of fault duration

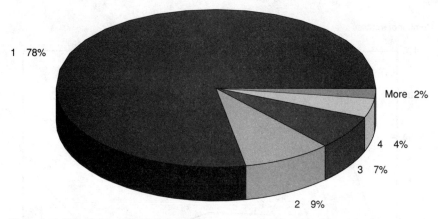

Figure 6.12 Distribution of fault referrals

Figures 6.13 through 6.18 analyze network operational expenditures. Figure 6.13 gives the overall scope of spending for hardware, software, infrastructure, communications, and human resources. As can be seen, one third is spent for the network management team ($150 million out of $450 million). This cost is significant, but not the highest in the industry. This ratio can approach 50 percent for human resources. Figures 6.14 through 6.18 display the distribution functions for each of the areas identified in Figure 6.13.

6.7 Summary

This chapter concentrated on the interpretation and consolidation of the data collected and presented in Chapter 5. For each area, standard and specific reports were prepared and presented. The companies are real but unspecified. Different companies contributed to the answers of the questionnaires, in-

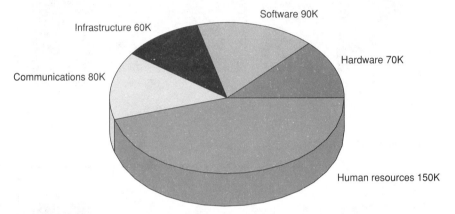

Figure 6.13 Distribution of expenditures for operations networks

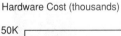

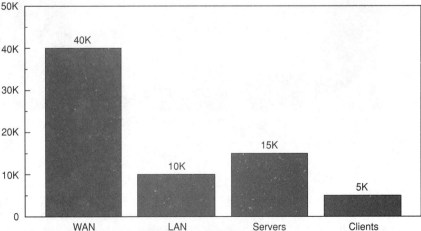

Figure 6.14 Distribution of hardware expenditures

Software Costs (thousands)

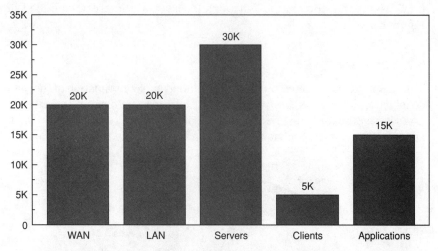

Figure 6.15 Distribution of software expenditures

Infrastructure Costs (thousands)

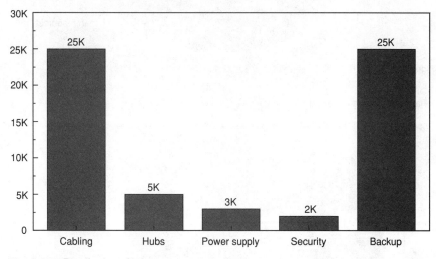

Figure 6.16 Distribution of infrastructure expenditures

depth functional analysis, and indicators. The typical benchmarking database for processors and networks usually contains significantly more samples. In the area of network management, however, benchmarking is relatively new. The comparative databases are still small, even in the case of leading benchmarking companies.

In this chapter, a flavor about reports is provided—both graphics and tables—that can be generated on the basis of information collected from previous chapters. Once the data have been consolidated and interpreted, benchmarking clients usually have very specific reporting needs. Assuming that the reporting tools are flexible, the client can generate very specific reports of their own. This chapter discusses descriptive indicators and reports only. The next chapter addresses comparative indicators and reports using industry averages.

Communications Costs (thousands)

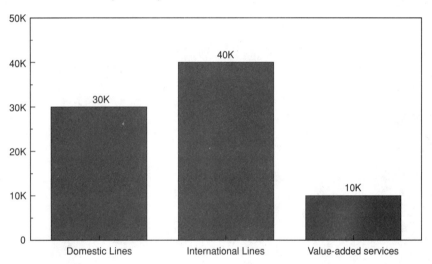

Figure 6.17 Distribution of communications expenditures

Human Resources costs (thousands)

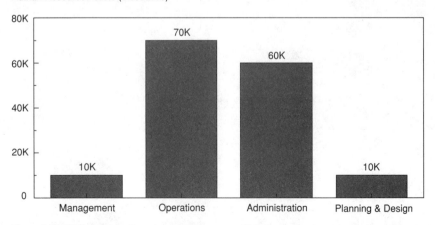

Figure 6.18 Distribution of human resources expenditures

7

Gap Analysis

Gap analysis is the process of identifying measures regarding network management functions, processes instruments, use of protocols, and the allocation of human resources to all of the above.

7.1 Introduction

Chapter 5 provided the raw data for data consolidation, analysis, and reporting. Chapter 6 concentrated on the interpretation of these data using descriptive indicators. All results have concerned single clients of the benchmarking service. Chapter 7 is different; here, the emphasis is on *comparative results*. Data are compared to the industry average, or the results from different companies are compared to each other. As far as possible, ideal models are shown to provide the ultimate targets for corporations. In most cases, however, the average of best practices is used as the more realistic target of improvement.

7.2 Model with Best Practices

To position companies, the benchmarking team needs targets and industry averages. Usually, the targets are organizations with over-average efficiency in operating their networks. To specify best practices, the forms from Chapter 3 are used.

Table 7.1 (Form 3.1) contains the entries for the ideal practice, which recommends 54 fully supported and 9 partially supported network management functions. In this case, it is assumed that each business area has its own supervisor headed by a network manager. This target is very difficult to reach. Usually, it takes several years to implement this level of functionality. The second version (Table 7.2) shows a more realistic version recommending 44 fully supported and 19 partially supported network management functions. In this case, fault tracking, change control, and fault monitoring are managed by the supervisors from other business areas. The supervisors report to the network manager.

TABLE 7.1 Support of Network Management Functions with Ideal Profile (Form 3.1)

	Fully supported	Partially supported	Not supported
Client Contact Point			
Receiving problem reports	x		
Handling calls	x		
Handling enquiries	x		
Receiving change requests	x		
Handling orders	x		
Making service requests	x		
Opening and referring trouble tickets	x		
Closing trouble tickets	x		
Operations Support			
Determining problems using trouble tickets	x		
Diagnosing problems	x		
Taking corrective actions	x		
Repairing and replacing	x		
Referencing to third-parties	x		
Backing up and reconfiguring	x		
Recovering	x		
Logging events and corrective actions	x		
Fault Tracking			
Tracking manually reported or monitored faults	x		
Tracking the progress and escalation of problems if necessary	x		
Distributing information	x		
Referring problems	x		
Change Control			
Managing, processing, and tracking service orders	x		
Routing service orders	x		
Supervising the handling of changes	x		

	Fully supported	Partially supported	Not supported
Planning and Design			
Analyzing needs	x		
Projecting application load		x	
Sizing resources		x	
Authorizing and tracking changes	x		
Raising purchase orders	x		
Producing implementation plans	x		
Establishing company standards		x	
Maintaining quality assurance		x	
Finance and Billing			
Asset management	x		
Costing services	x		
Client billing		x	
Usage and outage collection	x		
Calculation of rebates to clients		x	
Bill verification	x		
Software license control	x		
Implementation and Maintenance			
Implementing change requests and work orders	x		
Maintaining resources	x		
Inspection	x		
Maintaining configuration database		x	
Provisioning	x		
Fault Monitoring			
Monitoring the system and network for proactive problem detection	x		
Opening additional trouble tickets	x		
Referring trouble tickets	x		
Performance Monitoring			
Monitoring the system and networks performance	x		
Monitoring service-level agreements	x		
Monitoring third-party and vendor performance	x		
Optimizing, modeling, and tuning	x		
Reporting on usage statistics and trends	x		
Security Management			
Threat analysis	x		
Administration	x		
Detection	x		
Recovery	x		
Protecting the management systems	x		

TABLE 7.1 Support of Network ManagementFunctions with Ideal Profile (Continued) (Form 3.1)

	Fully supported	Partially supported	Not supported
Systems Administration			
Software version control	x		
Software distribution	x		
Systems management		x	
Administration of user - definable tables		x	
Local and remote configuration of resources		x	
Management of names and addresses	x		
Applications management	x		

TABLE 7.2 Support of Network Management Functions with Realistic Profile of Best Practices (Form 3.1)

	Fully supported	Partially supported	Not supported
Client Contact Point			
Receiving problem reports	x		
Handling calls	x		
Handling enquiries		x	
Receiving change requests	x		
Handling orders	x		
Making service requests	x		
Opening and referring trouble tickets	x		
Closing trouble tickets	x		
Operations Support			
Determining problems using trouble tickets	x		
Diagnosing problems	x		
Taking corrective actions	x		
Repairing and replacing	x		
Referencing to third-parties	x		
Backing up and reconfiguring	x		
Recovering	x		
Logging events and corrective actions	x		
Fault Tracking			
Tracking manually reported or monitored faults	x		
Tracking progress and escalation of problems if necessary	x		
Distributing information		x	
Referring problems	x		

	Fully supported	Partially supported	Not supported
Change Control			
Managing, processing, and tracking service orders	x		
Routing service orders	x		
Supervising the handling of changes	x		
Planning and Design			
Analyzing needs		x	
Projecting application load		x	
Sizing resources		x	
Authorizing and tracking changes	x		
Raising purchase orders	x		
Producing implementation plans	x		
Establishing company standards		x	
Maintaining quality assurance		x	
Finance and Billing			
Asset management	x		
Costing services	x		
Client billing		x	
Usage and outage collection		x	
Calculation of rebates to clients		x	
Bill verification		x	
Software license control	x		
Implementation and Maintenance			
Implementing change requests and work orders	x		
Maintaining resources	x		
Inspection		x	
Maintaining configuration database		x	
Provisioning	x		
Fault Monitoring			
Monitoring the system and network for proactive problem detection	x		
Opening additional trouble tickets	x		
Referring trouble tickets	x		
Performance Monitoring			
Monitoring the systems and networks performance	x		
Monitoring service-level agreements		x	
Monitoring third-party and vendor performance		x	
Optimizing, modeling, and tuning	x		
Reporting on usage statistics and trends	x		

TABLE 7.2 Support of Network Management Functions with Realistic Profile of Best Practices (Continued) (Form 3.1)

	Fully supported	Partially supported	Not supported
Security Management			
Threat analysis	x		
Administration	x		
Detection	x		
Recovery	x		
Protecting the management systems	x		
Systems Administration			
Software version control	x		
Software distribution	x		
Systems management		x	
Administration of user-definable tables		x	
Local and remote configuration of resources		x	
Management of names and addresses		x	
Applications management		x	

The ideal and practical versions of Form 3.2 are very similar to each other and are shown in Tables 7.3 and 7.4. Regarding the ownership of certain tools, absolutely no difference exists; the entries are identical. In both cases, integration is expected to happen with a combination of hierarchical management products and platforms. Element management systems are needed for all installed managed objects. Tables 7.3 and 7.4 show typical examples of the most likely used elements and their management systems. Full instrumentation is expected for monitors, analyzers, and security management tools. Slight differences can be identified with administration, the client contact point, and database instruments. The differences are based on timing and not on need for certain instruments.

TABLE 7.3 Network Management Instrumentation with Ideal Profile (Form 3.2)

Network management instruments	In use	Planned	Owner
Integrators			NM
Manager of managers	x		
Management platforms	x		
Element Management Systems for WANs			OS
Modems	x		
Multiplexers	x		
Packet switches	x		
Fast packet switches			

Network management instruments	In use	Planned	Owner
ATM switches			
SMDS nodes			
ISDN nodes			
Mobile communication			
Matrix switches			
Operations support systems			
Element management systems for LANs			OS
Bridges	x		
Routers	x		
Brouters			
Repeaters			
Extenders			
Hubs	x		
Segments			
FDDI			
DQDB			
Monitors and Analyzers			OS
			PM
WAN monitor	x		
LAN monitor	x		
WAN analyzer	x		
LAN analyzer	x		
Network monitor	x		
Software monitor	x		
Security Management Systems			SM
Protection of system and network	x		
Protection of management systems	x		
Administration Instruments			SA
Documentation systems	x		
Modeling instruments	x		
Presentation tools	x		
Report generators	x		
Trouble-tracking tools	x		
Software distribution tools	x		
Software licensing tools	x		
Database Tools			PM
Databases	x		
MIB browser	x		
Enquiry tools	x		
Client Service Point Instruments			CCP
Prediagnosis by phone		x	
Automated call distributor	x		
Voice mail	x		
E-mail	x		

TABLE 7.3   Network Management Instrumentation with Ideal Profile (Continued) (Form 3.2)

Network management instruments	In use	Planned	Owner
Pager	x		
Console emulator		x	
Expert system		x	

Key:
CCP = Client Contact Point
NM = Network Manager
OS = Operations Support
PM = Performance Monitoring
SA = Systems Administration
SM = Security Management

TABLE 7.4   Network Management Instrumentation with Realistic Profile of Best Practices (Form 3.2)

Network management instruments	In use	Planned	Owner
Integrators			NM
Manager of managers	x		
Management platforms	x		
Element Management Systems for WANs			OS
Modems	x		
Multiplexers	x		
Packet switches	x		
Fast packet switches			
ATM switches			
SMDS nodes			
ISDN nodes			
Mobile communication			
Matrix switches			
Operations support systems			
Element management systems for LANs			OS
Bridges	x		
Routers	x		
Brouters			
Repeaters			
Extenders			
Hubs	x		
Segments			
FDDI			
DQDB			
Monitors and Analyzers			OS PM
WAN monitor	x		
LAN monitor	x		
WAN analyzer	x		
LAN analyzer	x		

Network management instruments	In use	Planned	Owner
Network monitor	x		
Software monitor	x		
Security Management Systems			SM
Protection of system and network	x		
Protection of management systems	x		
Administration Instruments			SA
Documentation systems	x		
Modeling instruments		x	
Presentation tools	x		
Report generators	x		
Trouble-tracking tools	x		
Software distribution tools	x		
Software licensing tools		x	
Database Tools			PM
Databases	x		
MIB browser		x	
Enquiry tools		x	
Client Service Point Instruments			CCP
Prediagnosis by phone		x	
Automated call distributor		x	
Voice mail		x	
E-mail		x	
Pager	x		
Console emulator			
Expert system			

Key:
CCP = Client Contact Point
NM = Network Manager
OS = Operations Support
PM = Performance Monitoring
SA = Systems Administration
SM = Security Management

The use of network management protocols (Form 3.3) depends on the network architectures in use. Table 7.5 shows the ideal case, recommending the use of many protocols. Each of the recommended protocols can address a specific area. The model for the best practice (Table 7.6) concentrates on less-established protocols, but opens the future to the use of protocols. In both cases, the proprietary protocol entries are open; they depend on the architecture of the suppliers. For middle- and long-range, SNMPv1 will migrate to SNMPv2, and "Edge" might fully migrate to the OMNIPoint guidelines. Little importance is seen for CMIP over TCP/IP (CMOT) at this time.

TABLE 7.5 Network Management Protocols with Ideal Profile (Form 3.3)

Network management protocols	Established	Planned	Other protocols used
Proprietary protocols			
NMVT	x		
MSU			Depends on network
Dec			Depends on network
Novell			Depends on network
Others			Depends on network
CMIP	x		
SNMPv1	x		
SNMPv2	x		
CMOL	x		
CMOT			
XMP	x		
RMON	x		
OMNIPoint	x		
Edge	x		
Structured Query Language (SQL)	x		
Remote Procedure Calls (RPC)	x		

TABLE 7.6 Network Management Protocols with Realistic Profile of Best Practices (Form 3.3)

Network management protocols	Established	Planned	Other protocols used
Proprietary protocols			
NMVT	x		
MSU			Depends on network
Dec			Depends on network
Novell			Depends on network
Others			Depends on network
CMIP		x	
SNMPv1	x		
SNMPv2	x		
CMOL		x	
CMOT			
XMP		x	
RMON	x		
OMNIPoint		x	
Edge		x	
Structured Query Language (SQL)	x		
Remote Procedure Calls (RPC)		x	

In terms of human resources, the ideal model recommends establishing all job titles listed in Form 3.4. The realistic model representing the best practices (Table 7.7) compromises in 11 cases, recommending the planning of certain job titles without specific time limitations. Implicitly, it is assumed that new job titles can be established during organizational changes. In the realistic case, the organizational structure can appear as displayed in Figure 7.1.

To introduce powerful educational plans, skill requirements should be compared with existing skills and qualifying experiences. Table 7.8 shows the completed responsibility/skill-matrix of Form 3.5. This matrix represents the realistic view of best practices. In this case, it is not necessary to address the ideal model because it is identical to the realistic average of best practices.

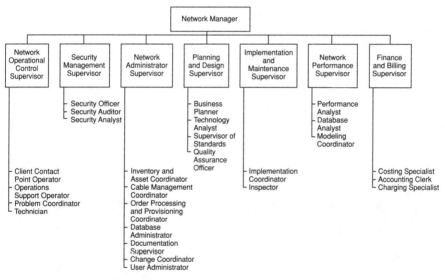

Figure 7.1 Organizational structure

TABLE 7.7 Human Resources Overview with Realistic Profile of Best Practices (Form 3.4)

Network management job titles	Established	Planned	Other job titles
Network manager	x		
Network administration supervisor	x		
Inventory and assets coordinator	x		
Cable management administrator	x		
Change coordinator	x		
Problem coordinator	x		
Order processing and provisioning coordinator		x	
Network operations control supervisor	x		
Client contact point operator	x		
Network operator	x		
Network technician	x		
Network performance supervisor	x		
Network performance analyst	x		
Database analyst	x		
Database administrator		x	
Modeling coordinator		x	
Security management supervisor	x		
Security officer	x		
Security auditor		x	
Security analyst		x	
Finance and control supervisor	x		
Costing specialist	x		
Accounting clerk		x	
Charging specialist		x	
Documentation supervisor	x		
Network maintenance supervisor	x		
Quality assurance officer		x	
Inspector		x	
Network implementation coordinator	x		
Service-level coordinator		x	
Supervisor of standards	x		
User administrator	x		
Design and planning supervisor	x		
Business planner		x	
Technology analyst		x	

TABLE 7.8 Responsibility/Skill Matrix with Realistic View of the Best Practices (Form 3.5)

Functions	Functional knowledge	General telecom	In-depth telecom	Instrument knowledge	Personal communication	Project management
Client Contact Point						
Receiving problem reports	X	X			X	
Handling calls	X			X	X	
Handling enquiries	X	X		X	X	
Receiving change requests	X	X			X	
Handling orders	X	X			X	
Making service requests	X	X			X	
Opening and referring trouble tickets	X	X	X			X
Closing trouble tickets	X	X		X		X
Operations Support Point						
Determining problems using trouble tickets	X	X	X	X	X	
Diagnosing problems	X	X	X	X	X	
Taking corrective actions	X	X	X	X	X	X
Repairing and replacing	X	X	X	X		X
Referencing to third-parties	X	X	X	X	X	X
Backing up and reconfiguring	X	X	X	X		X
Recovering	X	X	X	X	X	X
Logging events and corrective actions	X	X		X		

TABLE 7.8 Responsibility/skill Matrix with Realistic View of the Best Practices (Continued) (Form 3.5)

Fault Tracking

Functions	Functional knowledge	General telecom	In-depth telecom	Instrument knowledge	Personal communication	Project management
Tracking manually reported or monitored faults	x	x		x		x
Tracking progress and escalation of problems if necessary	x	x		x		x
Distributing information	x			x	x	
Referring problems	x	x	x	x	x	x

Change Control

Functions	Functional knowledge	General telecom	In-depth telecom	Business administration	Personal communication	Project management
Managing, processing, and tracking service orders	x	x		x	x	x
Routing service orders	x	x			x	x
Supervising the handling of changes	x	x			x	x
			Planning and Design			
Analyzing needs	x	x		x	x	x
Projecting application load	x	x		x	x	
Sizing resources	x	x	x		x	
Authorizing and tracking changes	x	x		x	x	x
Raising purchase orders	x			x	x	x
Producing implementation plans	x					x
Establishing company standards	x	x	x		x	
Maintaining quality assurance	x	x	x		x	

Finance and Billing

Functions	Functional knowledge	General telecom	In-depth telecom	Business administration	In-depth tariff information	Project management
Asset management	x	x				
Costing services	x			x	x	
Client billing	x			x	x	x
Usage and outage collection	x	x				
Calculation of rebates to clients	x			x		x
Bill verification	x	x		x	x	
Software license control	x					x

Implementation and Maintenance

Functions	Functional knowledge	General telecom	In-depth telecom	Practical experiences	Personal communication	Project management
Implementing change requests and work orders	x	x	x	x		x
Maintaining resources	x	x	x	x		x
Inspection	x	x	x	x	x	x
Maintaining configuration database	x	x	x	x		x
Provisioning	x	x	x	x	x	x

TABLE 7.8 Responsibility/Skill Matrix with Realistic View of the Best Practices (Continued) (Form 3.5)

Fault Monitoring

Functions	Functional knowledge	General telecom	In-depth telecom	Instruments knowledge	Personal communication	Project management
Monitoring system and network for proactive problem detection	x	x	x	x		
Opening additional trouble tickets	x	x		x		
Referring trouble tickets	x	x	x	x	x	x

Performance Monitoring

Functions	Functional knowledge	General telecom	In-depth telecom	In-depth instruments knowledge	Creativity and patience	Project management
Monitoring system and network performance	x	x	x	x		x
Monitoring service-level agreements	x	x	x	x		x
Monitoring third-party and vendor performance	x	x		x		
Optimizing, modeling, and tuning	x	x	x	x	x	x
Reporting on usage statistics and trends	x	x		x		x

Security Management

Functions	Functional knowledge	General telecom	In-depth telecom	Security issues	Creativity	Patience
Threat analysis	X	X	X	X	X	
Administration	X	X		X		X
Detection	X	X	X	X	X	X
Recovery	X	X	X	X		X
Protecting management systems	X	X	X	X	X	

Systems Administration

Functions	Functional knowledge	General telecom	In-depth telecom	Instruments knowledge	Personal communication	Project management
Software version control	X			X		X
Software contribution	X	X		X		X
Systems management	X	X		X		X
Administration of user-definable tables	X			X	X	
Local and remote configuration of resources	X	X	X	X		X
Management of names and addresses	X	X	X	X		
Applications management	X			X	X	X

7.3 Selected Indicator Examples for Building Industry Averages

This book cannot present a complete benchmarking database to build industry averages. Instead, based on Chapter 5, it offers simple examples to show how to build and use industry averages. Five sets of parameters exist, referenced as Companies 1 through 5; all are based on real numbers extracted from benchmarking studies, trouble tickets, logs, and monitoring results. The very basic computational steps have been omitted; only the results and industry averages are presented.

The real value of benchmarking companies is their database of benchmarking results. Such databases do exist for mainframe computers, servers, personal computers, some networking equipment, and well known networking architectures, such as SNA from IBM. Such databases cannot be purchased, but are populated from benchmarking experiences. The number of entries for benchmarking network management is still very low. It is expected that typical benchmarking companies, consulting companies, and systems integrators conduct benchmarks and maintain databases of results. Individual benchmarks can be conducted by various companies, but comparison with industry averages is done by only a few companies.

Industry averages are expected to be built for the following industries:

- Retail
- Transportation
- Finance
- Manufacturing
- Government
- Military

Cross-industry averages consider all these industries. To build and maintain the network management benchmarking database, the items discussed in the following subsections are important.

Screening the companies before adding the data to the industry average. Sometimes, very special solutions for network management exist because of certain networking architectures or applications, which would move the industry average into areas where statistical representance can no longer be guaranteed.

Authorization by client to include private data in the database. Agreement from companies is likely, but company identification must not be revealed in the database. At the same time, however, the benchmarking company must be aware of the company to update data and build industry averages.

Updating cycles for industry averages. The benchmarking company should decide how frequently the industry averages should be recomputed. Alternatives include annually or periodically after finishing each benchmarking project.

Weighting companies. The benchmarking company must decide whether weighting is necessary after successfully conducting benchmarking projects. Plain averaging does not work because smaller companies with exotic solutions would move the average into nonrealistic directions. Benchmarking companies must consider weight on the basis of the number of managed objects.

Selection of database. Because of the unexpected enquiries and the demand of special evaluations, a relational database seems to be the right choice. The product is not as important; five to six products can satisfy the needs of the benchmarking company.

Applications. Applications include a range of statistical packages, reporting software, and graphics for visualization of results. In this respect, no fundamental differences to the tools listed exist for supporting general benchmarking.

Generic indicators

The industry average shows the following results for the quality of attended operations for the five companies:

Outstanding	1.2
Good	4.2
Fair	1.2
Poor	1.4

It is clear that *good* dominates. Figure 7.2 shows this result.

Organizational indicators

To compare various businesses, it is important to compare the ratio between number of managers, subject-matter experts (SME), and external consultants. The industry average shows the following results for the referenced companies:

Manager/SME-ratio without external resources	18.4%
Manager/SME-ratio with external resources	17.0%

The difference between the two values is not significant. The average is in the high range, considering an operational range between 10 and 20 percent.

Number of companies

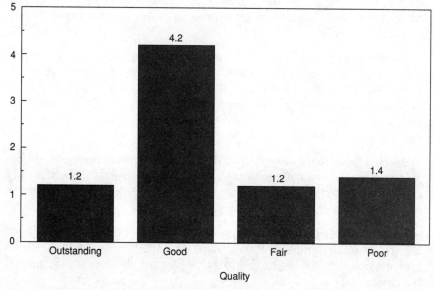

Figure 7.2 Industry average for quality of attended operations

Indicators for specific network management processes

To characterize the efficiency of the client contact point, the indicator of handling troubles over the phone is used. The calculated industry average is the following:

Average delay to answering calls	8.6 sec.
Maximum delay to answering calls	33 sec.

Similar calculations can be made for all other indicators and communication forms, such as voice mail and E-mail.

The grade of service at the client contact point is measured quantifying the response quality to various problems. The industry average is the following:

Outstanding	1.6
Good	3.6
Fair	2.0
Poor	0.8

It is clear that *good* dominates for the selected five companies; *fair* and *outstanding* balance each other. Figure 7.3 shows this result graphically.

The quality of change management is measured by two indicators: completion rate for change requests and successful execution of changes. Both numbers relate to the total number of change requests for a finite period of time. The industry average for the semiannual timeframe is the following:

Completion ratio	97.6%
Successful execution ratio	97.4%

Both are very ambitious results not easy to reach, particularly for the completion ratio of change requests forms.

The duration of changes can become extremely crucial to the efficiency of operations. The industry average has been computed for top severity changes. Figure 7.4 shows the frequency distribution of the changes over the selected time windows on the basis of the entries in section 5.4. The form of the density function is an expected one; the peak is at 8 hours of change duration.

In reengineering the change management process, the number of change management steps is important. The industry average shows the following results for this indicator:

Persons involved in changes	7
Steps in the change management process	12
Automated steps in the change management process	2
Manual steps in change management process	8.6
Semiautomated steps in change management process	1.2

The result is realistic; 72 percent of the change management process steps are still manual.

Number of Companies

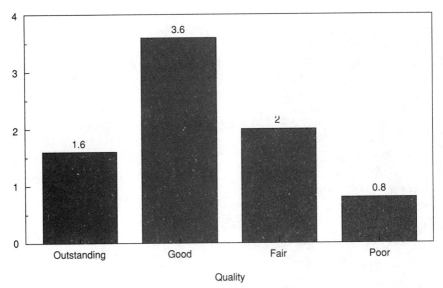

Figure 7.3 Industry average for grade of service at client contact point

Percent of problems resolved

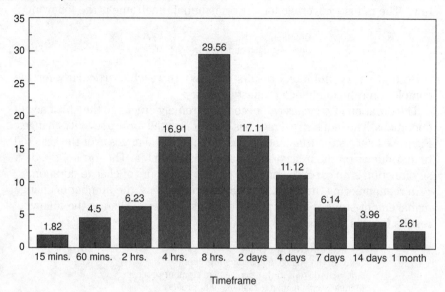

Figure 7.4 Frequency distribution for duration of top severity changes

Problem resolution time can substantially be reduced by recognizing problems early. Proactive monitoring is very helpful in this respect. Based on proactive monitoring results, the industry average has been calculated for a time window of one quarter. The value is 20.2 percent, which is unexpectedly low. Corporations should target much higher values in the range of 50 to 60 percent

The duration of troubles heavily impacts the service quality of the whole organization. The distribution of the duration helps make the right investments into instrumentation and human resources. Figure 7.5 shows the density function of trouble resolution for tail circuits. The form of the density function is an expected one; the peak is for a resolution duration of one hour.

The speed of problem resolution is heavily impacted by the number of ownership changes, called *referrals*. The more referrals, the longer the fault resolution process takes. Based on 30 days in five companies, the industry average shows the following results:

No. of referrals	Percent
1	72.80
2	11.92
3	6.73
4	6.18
More than 4	2.80

Figure 7.6 shows these results. The general form of the chart is typical for one and two referrals. The stagnation for three and four referrals comes randomly from the samples taken.

Availability is the most popular indicator of service-level agreements between the users and the service providers. The industry average for networking facilities is 98.43 percent. This number is good and close to the realistic mark of 99.5 percent availability for networking facilities.

Utilization data for applications, equipment, and facilities are extremely valuable for determining future upgrades and capacity extensions. Tables 7.9 and 7.10 show two targets. Table 7.9 shows the ideal model requesting availability of data for all resources under consideration, high resolution rates,

Percent of problems resolved

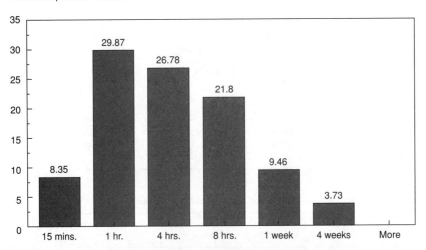

Figure 7.5 Frequency distribution for trouble resolution of tail circuits

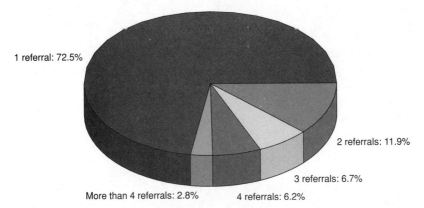

Figure 7.6 Distribution of trouble referrals

reasonable consolidation timeframes, and exclusively monitoring technology for data collection. Table 7.10 considers present status, the availability of monitoring technology, and practical limits of data volumes. As a result, Table 7.10 shows compromises for targeted availability of utilization data.

TABLE 7.9   Availability of Utilization Data—Ideal Model

Indicators	Data	Level of details	Consolidation periodicity	Instrument man. mon.
Applications				
Batch	yes	hr.	day	yes
Online	yes	min.	hr., day	yes
Mail	yes	hr.	day	yes
Equipment				
Server	yes	min.	hr., day	yes
Client	yes	min.	hr., day	yes
Router	yes	min.	hr., day	yes
Multiplexer	yes	min.	hr., day	yes
Facilities				
WAN	yes	min.	min., hr., day	yes
MAN	yes	sec.	min., hr., day	yes
LAN	yes	sec.	min., hr., day	yes

TABLE 7.10   Availability of Utilization Data—Realistic Model

Indicators	Data	Level of details	Consolidation periodicity	Instrument man. mon.
Applications				
Batch	yes	day	day	yes
Online	yes	hr.	hr., day	yes
Mail	yes	hr.	day	yes
Equipment				
Server	yes	hr.	hr., day	yes
Client	no			
Router	yes	min.	hr., day	yes
Multiplexer	yes	min.	hr., day	yes
Facilities				
WAN	yes	min.	hr., day	yes
MAN	no			
LAN	yes	min.	min., hr., day	yes

Cost indicators

Costing is a must, and charging is an option for accounting with networking services. In most cases, five categories are identified for grouping cost items. The industry average shows the following results for those items:

Hardware	14.13%
Software	20.34%
Infrastructure	11.27%
Communications	19.60%
Human resources	34.74%

Figure 7.7 depicts these results graphically. The dominating component is human resources, consuming approximately 45 percent of costs.

7.4 Positioning Companies Using Comparative Indicators

Using experiences of best practices, targeted indicator ranges, and industry averages, companies can position themselves or the benchmarking team can identify gaps by comparing the performance of different companies. This segment gives a number of examples for a variety of comparisons.

Using quantifiable indicators for comparisons

Quantifiable indicators include generic, organizational, those for specific network management processes, and cost. Each are described in the following subsections.

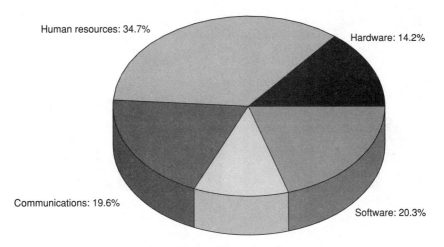

Figure 7.7 Distribution of networking expenditures

Generic indicators. Figure 7.8 shows the performance of attended operations of Company 1 in comparison with the industry average. The form is identical, and differences are minimal. The conclusion is that Company 1 performs equivalent to industry average.

Organizational indicators. The manager-to-subject-matter-expert ratio is displayed in Figure 7.9. Industry average is also indicated. All five companies are in the same range and very close to the industry average. Acceptable range is between 10 and 20 percent.

Specific network management process indicators. One of the indicators for the efficiency of the client contact point is how quickly trouble calls are answered. Figure 7.10 shows the average and maximal delays for phone calls. The industry average is shown for both indicators to orient the company under consideration.

Figure 7.11 shows the grade of service at the client contact point of Company 2 compared to the industry average. The form is identical, the only difference is that Company 2 does not have the *poor* entry for any of the problem categories supported by the client contact point.

Comparing Company 3 with the industry average for quality of change management indicates that Company 3 operates slightly above the industry average:

Completion ratio for change requests	0.40% (98.00%–97.60%)
Successful execution ratio	0.60% (98.00%–97.40%)

Number of companies

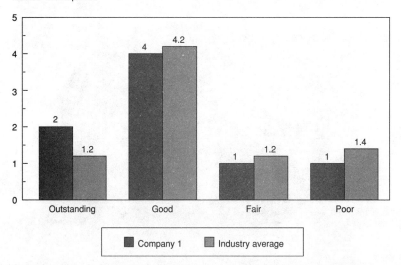

Figure 7.8 Attended operations—Company 1 compared to industry average

%

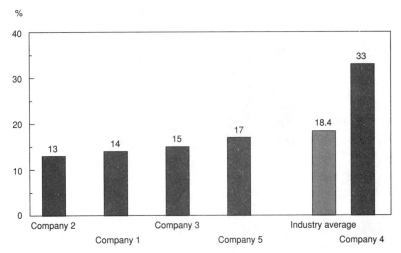

Figure 7.9 Manager-to-SME ratio for five companies and comparison to industry average

(A) Average response delay
Seconds

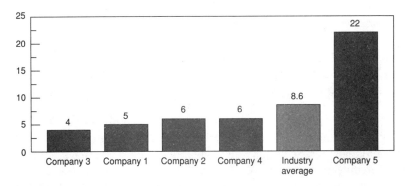

(B) Maximal response delay
Seconds

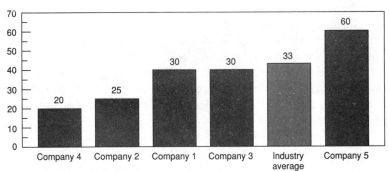

Figure 7.10 Handling trouble calls

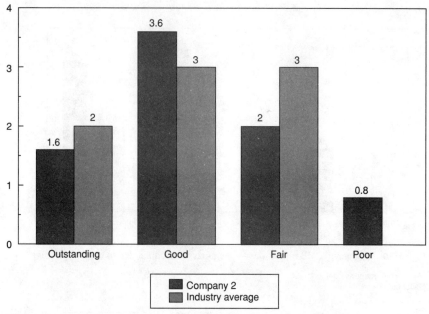

Figure 7.11 Grade of service—Company 2 compared to industry average

The goal should be to remain at this leading edge of the industry.

Figure 7.12 illustrates three density functions: industry average, Company 2, and Company 5 for the duration of severity 1 changes. The forms are similar with slight differences in certain time windows.

The duration of change management depends on the number of process steps. Figure 7.13 shows the industry average and the number of steps for all five companies. As the industry average indicates, the manual steps still dominate change management.

A similar presentation form has been selected for the proactive fault detection ratio. Figure 7.14 displays the results for all five companies; the industry average is also indicated.

Figure 7.15 illustrates three density functions for the duration of troubleshooting tail circuits: the industry average, Company 2, and Company 3. The forms are similar with slight differences in certain time windows.

Trouble referrals are important to evaluate the efficiency and skill level of tiers that deal with issues in various phases of trouble resolution. To compare the numbers, Table 7.11 displays the results for all five companies, including the industry average.

Part of the service-level agreements deal with end-user availability. In particular, in telecommunications, the offered and confirmed availability can significantly influence the fate of bids. Availability of facilities is very important in the networking business. Figure 7.16 displays the average availability for facilities for all five companies; the industry average is included as well.

Number of changes

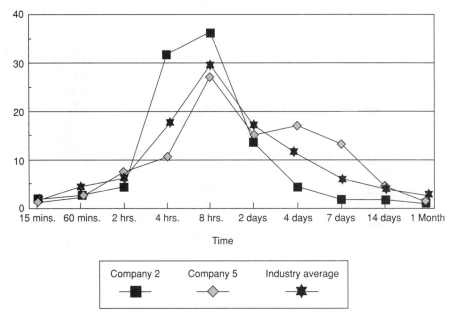

Figure 7.12 Comparison of density functions for duration of top severity changes

Number of
change process steps

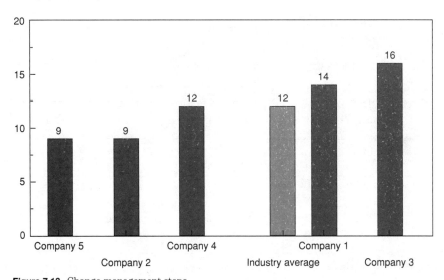

Figure 7.13 Change management steps

Percent of
faults detected

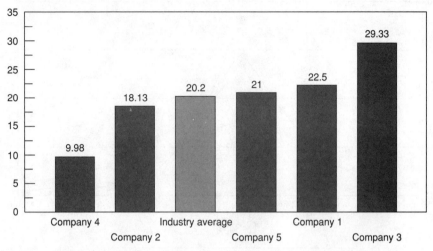

Figure 7.14 Proactive fault detection ratio

Number of troubles

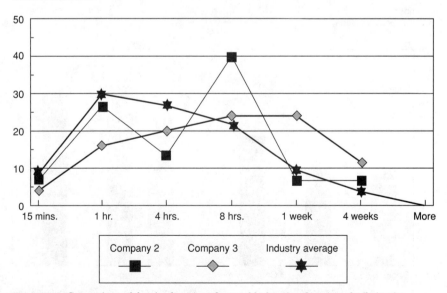

Figure 7.15 Comparison of density functions for troubleshooting duration of tail circuits

Number of facilities

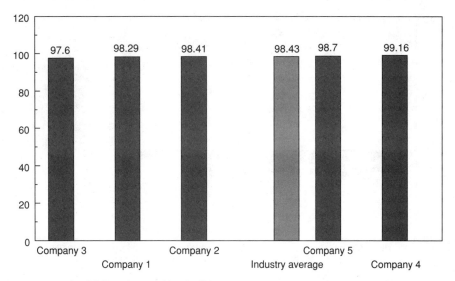

Figure 7.16 Availability of networking facilities

TABLE 7.11 Summary of Fault Referrals

Companies	1%	2%	3%	4%	>4%
Company 5	89.56	6.70	4.52	2.16	0.65
Company 3	89.21	4.70	2.40	2.57	0.77
Company 2	78.00	9.00	7.00	4.00	2.00
Industry average	*72.80*	*11.92*	*6.73*	*6.18*	*2.80*
Company 1	70.86	18.25	6.04	2.18	2.67
Company 4	36.39	20.94	13.71	19.90	7.90

Costs. This comparison is usually the highest priority level. Not only the size of the budget for communications and network management but also the allocation of money is very interesting to corporations. Figure 7.17 compares the distribution of spending for Companies 1 and 2 with the industry average.

For hardware, software, and infrastructure, both companies spend less than industry average. Spending for communications is different; Company 1 spends more, Company 2 spends less. Both companies spend considerably more for human resources than industry average indicates.

Positioning using forms

Positioning the company means that Forms 3.1 through 3.5 are used to compare the entries of the ideal model against the realistic average of best

Industry average

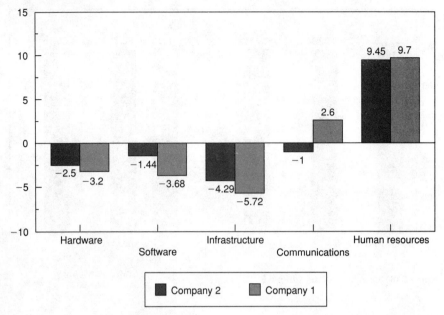

Figure 7.17 Comparison of expenditures

practices. The results for Company 2 were presented in section 5.6. The desired entries for the ideal and realistic models have been addressed in section 7.2. The positioning using these forms shows the following results for Company 2.

Support of network management functions. Table 7.12 displays all network management business areas and evaluates how individual network management functions are supported. Both the ideal and realistic models are considered.

Supporting network management by instruments. Table 7.13 shows the support of network management functions by instruments. All major instrument groups are evaluated against the ideal and realistic models.

Support of network management protocols. In both cases, just one third of the protocols are supported, but the key protocols are among them. A strong indication is that SNMPv1 will be extended to SNMPv2 and message service unit (MSU) will one day fully replace network management vector transport (NMVT).

TABLE 7.12 Support of Network Management Functions

Models	Match with ideal model (%)	Match with average of best practices (%)
Client contact point	50.00	62.50
Operations support	62.50	62.50
Fault tracking	50.00	75.00
Change control	33.33	33.33
Client contact point	50.00	62.50
Planning and design	25.00	37.50
Finance and billing	14.30	28.60
Implementation and maintenance	40.00	60.00
Fault monitoring	66.66	66.66
Performance monitoring	40.00	60.00
Security management	40.00	40.00
Systems administration	14.30	42.90

TABLE 7.13 Support of Network Management Functions by Instrument

Models	Match with ideal model (%)	Match with average of best practices (%)
Integrators	50.00	50.00
Element management systems (WAN)	66.66	66.66
Element management systems (LAN)	100.00	100.00
Monitors and analyzers	83.33	83.33
Security management tools	50.00	50.00
Administration tools	42.80	57.10
Databases	33.33	33.33
Client contact point tools	16.66	16.66

Support of job titles for network management. Serious gaps exist in this area. The matching ratio with the ideal model is as low as 35.30 percent. Even with the average of best practices, the match ratio is not higher than 50 percent. The reason could lie with the traditional method of organizing network support. The average organization definitely needs two to three iterations until the desired level is reached.

Analysis of skill levels. Analysis of skill levels is a very difficult area because subjectivity can influence the final results. Table 7.14 shows how existing skill levels match expectations for each network management functional group. As can be seen, much room for improvement exists.

TABLE 7.14 Analysis of Skill Levels

Functions / Number of matches	Functional knowledge	General telecom	In-depth telecom	Instrument knowledge	Personal communication	Project management
Client Contact Point	5	2	0	4	0	0
Operations Support	8	1	7	8	2	0
Fault Tracking	4	2	1	1	1	0

Functions / Number of matches	Functional knowledge	General telecom	In-depth telecom	Business administration	Personal communication	Project management
Change Control	3	3	0	3	1	1
Planning and Design	3	3	0	1	0	0

Function / Number of matches	Functional knowledge	General telecom	In-depth telecom	Business administration	In-depth tariff information	Project management
Finance and Billing	3	0	0	2	0	1

Function / Number of matches	Functional knowledge	General telecom	In-depth telecom	Practical experience	Personal communication	Project management
Implementation and Maintenance	5	0	4	3	0	0

Gap Analysis 273

Function	Functional knowledge	General telecom	In-depth telecom	Instruments knowledge	Personal communication	Project management
Number of matches						
Fault Monitoring	3	2	1	3	0	0

Function	Functional knowledge	General telecom	In-depth telecom	instruments knowledge	Creativity and patience	Project management
Number of matches						
Performance Monitoring	3	2	1	1	1	0

Function	Functional knowledge	General telecom	In-depth telecom	Security issues	Creativity	Patience
Number of matches						
Security Management	4	0	3	1	1	

Function	Functional knowledge	General telecom	In-depth telecom	Instruments knowledge	Personal communication	Project management
Number of matches						
Systems Administration	4	2	0	1	1	1

7.5 Summary

Chapter 7 provided quantitative information on the ultimate targets of performance for corporations. This target was based on the so-called ideal model. A more realistic basis is when the averages of best practices are considered targets. This target is based on the realistic model. The benchmarking team can compare consolidated data and results from Chapter 6 with the ideal or with the more realistic model for multiple companies, even within the same company if multiple powerful business units compete against each other. Chapter 7 has shown examples for all of these cases. The more realistic model usually works with industry averages. Averages are built and updated with life data from various environments. The real power of benchmarking companies can be measured by the number of entries in their benchmarking database. In the case of network management, the databases are not yet large enough. Most of the graphics and tables of this chapter can be used during the final presentation of results to top management.

Outsourcing Network Management

It is not simple to help users improve network management and networking efficiency. Network-management services offered by mainframe vendors and network suppliers can also help offload users in all or some network management areas. The alternative offerings are very broad at the moment. They start with opening a window into the physical network of the supplier for users to provide additional physical network management data for correlation by clients. These choices are offered by INMS (MCI), Accumaster Management Services (AT&T), and Concert from British Telecom. In other words, segments of data or applications are offered. In another case, network management subsystems are completely supported on-premise or remotely using special management instruments of the supplier. The ultimate alternative is to outsource network management completely to third parties, leaving the overall supervisory and control functions and a shell of network management functions with users. The suppliers are industry giants, like IBM, DEC, and AT&T, or service companies, such as Electronic Data Systems (EDS), MCI, U.S. Sprint, Eunetcom, Unisource, RBOCs, Loral or smaller start-up companies, such as MAXM, I-Net, or Network Management, Inc.

8.1 Basic Outsourcing Decisions

Which services to take and from whom are difficult questions to answer. Table 8.1 helps identify which network-management business areas can be

TABLE 8.1 Outsourcing of Network Management Functions

Business areas	Outsourcing Decision		
	Yes	No	Uncertain
Client contact point			x
Operations support	x		
Performance monitoring			x
Security management		x	
Finance and billing	x		
Change control			x
Implementation and maintenance	x		
Fault tracking			x
Network planning and design		x	
Fault monitoring	x		
Systems administration	x		

considered for outsourcing. Outsourcing is likely for operations support, finance and billing, implementation and maintenance, fault monitoring, and systems administration; it is not likely for network design and security management; and it is uncertain for client contact point, performance and management, change control, and fault tracking.

Before deciding for or against outsourcing, the following criteria must be evaluated very carefully (TERP91A):

1. What are the present costs of network management equipment, communications, and people?

2. What is the full visibility of existing processes, instruments, and human resources to decide which functions are to be considered for outsourcing? Outsourcing is a good excuse to audit present operations and address areas that need improvement. The result of preparing outsourcing could be insourcing.

   Internal analysis by internal or external analysts can result in substantial savings in operating expenses (30% to 40%), staff reduction (25% to 50%), and stabilizing network budgets.

3. What is the network availability, indicating the highest level of risks the customer must include into the service contract? Many times, certain vendors fall short at the very beginning by not being able to guarantee the targeted availability.

4. What is the grade of service required by users and applications, which can dictate a certain type of outsourcing company that does not share its resources among multiple clients?

5. What are the security standards and tolerable risks that prohibit third-party vendors from gaining access to the network and its carried traffic?

6. What is the concentration on the business? This concentration could require that the operation not build a sophisticated network manage-

ment system or organization, but rather concentrate on the technology of the business.

7. What is the availability of network-management instruments? If the company invested substantial amounts into instrumentation, outsourcing should be favored; if not, outsourcing can still be considered, but with a lower priority.

8. What is the availability of skilled network management personnel? Most frequently, this issue can be the only driving factor for outsourcing. Not only present status, but future needs and their satisfaction must be quantified prior to outsourcing.

9. What is the stability and growth rates of the environment? These issues have serious impacts on the contract with the vendor. Acquisitions, mergers, business unit sales, and application portfolio changes need special and careful treatment in contracts.

10. Will the company offer value-added services to other third parties? In certain corporations, if bandwidth of communication resources is underutilized, it can be used for offering "low-priority" services to third parties who cannot afford to build a network on their own (such as point-of-sale applications).

11. Can the company construct good outsourcing contracts? The length of outsourcing contracts—often of 7 to 10 years—means that the wording of the contract is extremely important.

12. What is the philosophy of network management, including horizontal and vertical integration, centralization, automation, and the use of a network-management repository? These details should be in concert with the offer and capabilities of the outsourcer.

To answer what type of service partner to use, Table 8.2 offers the pros and cons of various types. The final decision depends on the networking environment and budgets for third-party network management. Summarizing the expectations, outsourcers are expected to meet the following requirements:

- Financial strength and stability over a long period of time
- Proven experience in managing domestic and multinational networks
- Powerful pool of skilled personnel
- Tailored network-management instruments that can be used exclusively or in shared modes for the clients under consideration
- Proven ability in implementing the most advanced technology
- Outstanding reputation in conducting business
- Willingness for revenue sharing
- Fair employee transfers

TABLE 8.2 Service Provider Evaluation

Vendor type	Pros	Cons
Mainframe vendor	■ Knowledge of data networks ■ Reputation ■ Knowledge of logical network management ■ Vertical integration	■ Dependency on proprietary architecture ■ Data orientation ■ Horizontal integration ■ Physical network management
Carrier	■ Knowledge of voice network ■ Reputation ■ Knowledge of physical network ■ Horizontal integration	■ Dependency on proprietary architecture ■ Vertical integration ■ Logical network management ■ Voice integration
Integrator	■ Knowledge of both data and voice networks ■ Knowledge of both vertical and horizontal integration ■ Knowledge of both logical and physical integration ■ Actual implementation	■ High-cost ■ Lack of reputation ■ Project-driven
Consultant	■ Knowledge of both data and voice networks ■ Knowledge of both vertical and horizontal integration ■ Knowledge of both logical and physical integration ■ Leading edge of technolgoy	■ High-cost ■ Lack of reputation ■ Lack of actual implementation experience

The weights of the criteria must be set by the client during evaluation.

Multiple choices exist for the service contract, including on-premise, off-premise, or simple monitoring functions. Table 8.3 compares these three choices.

The fate of human resources is a difficult question. Many times, the cost of personnel is what forces people to consider outsourcing. It is very difficult to estimate the most likely path that people's careers take. From today's perspective, 20 percent of staff can be cross-educated and kept; this number could increase to 30 percent in the future. Another 50 percent of staff can be taken over by the network management service provider, which will peak at approximately 60 percent in the future. Finally, 30 percent of staff can be laid off, although the trend for this figure is estimated to decrease to approximately 10 percent.

8.2 Outsourcing Agreements

In establishing an agreement, the customer is made as responsible as the outsourcers for ensuring that the outsourced services are supplied adequately. As the outsourcer should be promoting business goals, the service levels identified in the contract should support these goals (TERP, 1992). Benchmarking results can help justify outsourcing decisions. They highlight expenses and gaps in supporting network management functions, the use of network management instruments, and the allocation of human resources to functions and instruments.

Prior to signing an outsourcing contract, the items discussed in the following subsections should be addressed and agreed upon.

Who is authorized to modify the contract?   The provisions for modifying the agreement should appear somewhere in the agreement, preferable in the beginning. Basically, the provisions should allow for either the outsourcer or the customer to reopen negotiations and for management approval if the priority of the work is to be changed. Generally, changes should be made only after an analysis has demonstrated that the problem is not an aberration. Trial agreements for new outsourcing services could be a meaningful approach.

Duration of agreements.   The agreement should be written as ending after a certain period of time, such as a certain number of years. Alternatively, all outsourcing agreements might be scheduled for revision after expected changes to hardware, networking nodes, facilities, or software. Above all, no one should believe that the agreement is a commitment for an eternity, regardless of changes in the business or networking environment.

Reviews.   Reviews are necessary for considering any impacts of this dynamically changing environment. For mutual benefits, any impacts must be openly discussed and the necessary changes written into the existing contract.

TABLE 8.3 Service Alternatives

Type of service contract	Pros	Cons
On-site (on-premise)	■ Rapid troubleshooting ■ Dedicated personnel ■ Continuous consultation	■ High cost ■ No shared instrumentation ■ Space requirements
Monitoring (local, remote, or both)	■ On-demand activity ■ Indicators selected by users and suppliers ■ Shared instrumentation	■ Fragmented network management ■ No dedicated personnel
Off-site (off-premise)	■ Full scope of network management ■ Shared instrumentation ■ Combination with virtual networking services	■ Moderate cost ■ No dedicated personnel ■ Longer troubleshooting ■ No continuous consultations

Service-level indicators. Service can be agreed for various levels of detail, including on-premise monitoring and off-premise solutions. Principal service indicators should include the following:

- Availability of networking components, such as facilities and equipment, with a differentiation by applications preferable
- Response time segmented by users, applications, devices, and communication forms (data, voice, image, and video)
- Grade of service, including physical parameters of network facilities
- Distribution of faults identifying short and long outages
- Mean time between failures (MTBF)
- Mean time to repair (MTTR)

User commitments. User commitments include informing the outsourcers about lines of business, strategic goals, critical success factors, networking environments, organizational changes, application portfolios, service expectations and indicators, directions of technology in networks design, and any early warnings of the need to renegotiate the contract if necessary.

Reporting periods. Performance reports on key service indicators should be regular, such as weekly or monthly, and copies should go to the organization entity in charge, as well as information technology (IT) and information systems (IS) management. Reports need a format agreed upon by both parties. Performance reports provide an opportunity to identify impending problems and propose solutions in advance. They are key tools for avoiding crisis management.

Costs and chargeback policy. Contracting parties must agree on the conditions of payments, chargeback reports, alternatives to bill verification, and expected inflation rates. The transfer of human resources must also be negotiated and agreed upon.

Penalties for noncompliance. Creating penalties for noncompliance is relatively simple with an appropriate costing and chargeback policy. For noncompliance of service objectives, payments to outsourcers should be reduced. In most cases, the damage caused by noncompliance is much more severe than the penalty reimbursed in monetary units.

Employee transition. Somewhere in the agreement, usually in the closing section, employee transitions must be addressed. This section includes the names of employees transferred to the outsourcers and the conditions of takeover, such as salaries, job security, title, and position. Training and education that will be provided must also be identified.

Billing and currency issues. It is beneficial to keep payments to a single currency whether payments are made to a central contact or dispersed. This issue is of particular importance for multinational companies.

Periodic pricing reviews. Subcontractors and telecommunication tariffs lower or raise costs over time. The contract must observe these changes and be adjusted, if necessary.

Outsourcing contracts will migrate to a relatively stable number in the future. Total outsourcing will be rare in the future, with partial outsourcing more popular. Partial outsourcing could be renamed as *strategic partnerships* between the outsourcers and the outsourcing companies.

8.3 Telecommunications Outsourcing

Telecommunication providers are receiving a lot of attention from companies who want to outsource. In particular, multinational companies should outsource at least the backbone segment of their networks. Most outsourcers provide a single point of contact (SPOC), offloading the companies from dealing with various Post, Telegraph, and Telephone companies (PTTs), private service providers, and local phone companies for international billing, order processing, and provisioning.

To meet clients' needs of faster response times and a better quality, telecommunication providers form global alliances. In this case, the client receives the outsourcing service from the partner who represents the alliance in the specific geographical area. Behind the scenes, however, many other suppliers help provision the service. Table 8.4 (YANK94) shows current global alliances of telecommunication providers. It is a very dynamic area with many changes to come.

WordPartners is targeting global frame relay, global virtual networks, global private lines, and global 800-services in addition to outsourcing. In particular, the Unisource partnership plans to penetrate multinational companies with a strong European presence. The shareholders of this alliance are the following:

AT&T	40%
KDD	24%
Singapore Telecom	16%
Unisource	20%

TABLE 8.4 Global Alliances

Company	Alliances					
	WorldPartners	Unisource	Concert	Infonet	Eunetcom	Cable & wireless
AT&T	x					
Belgacom				x		
British Telecom			x			
Deutsche Telekom				x	x	
France Telecom				x	x	
Hong Kong Telecom	x					
KDD (Japan)	x			x		
Korea Telecom	x					
MCI			x			
Mercury						x
Bermuda						x
CWI						x
Caribbean						x
Hong Kong Telecom						x
IDC						x
Optus						x
Tele 2						x
PTT Netherlands		x		x		
Schweizer PTT		x		x		
Singapore Telecom	x			x		
Telefonica		x		x		
Telia (Sweden)		x		x		
Telstra (Australia)	x			x		
Unitel (Canada)	x					
USSprint					x	

Unitel, Telstra, Telecom New Zealand, and Hong Kong Telecom are partners but have no shareholders—yet. Unisource keeps very tight connections to Sita, taking the responsibility of operating the global Sita network in multiple countries.

British Telecom dominates the Concert alliance with 75 percent. MCI holds the other 25 percent. British Telecom, however, has acquired 20 percent of MCI. Concert targets global network outsourcing, frame relay, virtual networks, and X.25 and managed private lines.

The alliance between Sprint and Atlas/Eunetcom targets global virtual networks, X.25, private lines, and calling cards. Atlas/Eunetcom is a special equity alliance between France Telecom and German Telekom. The whole alliance embeds France Telecom and German Telekom each with 10 percent, with 80 percent controlled by Sprint, shared between backbone offers (30 percent) and NAFTA sales units (50 percent).

Infonet has many owners. The ownership picture is extremely complex with the following distribution:

Belgacom	7.17%
France Telecom	21.56%
Deutsche Telekom	21.56%
PT T Netherlands	7.17%
Telia	7.17%
Schweizer PT T	7.17%
Telefonica	7.17%
Singapore Telecom	7.17%
Telstra	7.17%
KDD (Japan)	6.67%

Atlas/Eunetcom reaches 43.12 percent, WorldPartners 21.01 percent, and Unisource 28.68 percent. When the new deal between WorldPartners and Unisource is approved by governments, the new alliance will be slightly larger than Atlas/Eunetcom (by 6.57 percent). Infonet offers EDNS, MNDS, InfoLAN, global frame relay, and global X.25 services besides outsourcing.

Cable & Wireless dominates three major geographical areas: Asia (Hong Kong Telecom, IDC, Optus), the Americas (CWI, Bermuda, the Caribbean), and Europe (Mercury and Tele2 from Sweden). The service offering is manifold, including dedicated digital circuits, managed private lines, splitstream route diversity, global X.25, frame relay, and SNA services.

IBM might earn a small size of the global pie with Advantis, its alliance with Sears.

8.4 Summary

Outsourcing is being seriously considered by corporations to better control network management related expenditures. Because of revolutionary changing telecommunication technology, the ever-increasing user demands for new services, and the shortage on skilled network management staff drive businesses to evaluate outsourcing alternatives. This evaluation must be quantitative and in-depth and will take time. Benchmarking indicators and a comparison with best practices help decide on outsourcing or insourcing.

9

Recommendations, Reports, and Final Results

The final product is the benchmarking report to be presented to various management layers within the benchmarked company.

9.1 Introduction

The final results of benchmarking is to present specific recommendations, which consist of the descriptive and comparative indicators and any recommendations about outsourcing. In addition, generic recommendations are included about performing certain activities while avoiding others during implementation. This generic segment is based on past experiences of the benchmarking team. Usually, not all recommendations apply to a specific company.

To improve overall performance, very concrete recommendations are expected in the areas of specific indicators and outsourcing. These recommendations should include the project plan, reasonable schedules, and the expected human resources and equipment needed.

9.2 Specific Recommendations

Specific recommendations include descriptive indicators and comparative indications. Both are described in the following subsections.

Descriptive indicators

Descriptive indicators include the following:

- Network management functions by business areas
- Network management processes
- Specific indicators

Network management functions by business areas. In this segment, 19 network management functions have been analyzed in depth. Improvements are expected in the areas described in the following subsections.

Increase the automation level for receiving problem reports from users. The dialog with end users is always time-consuming. Users cannot describe their problems accurately, they misjudge the nature of symptoms, and they delay their responses. All these factors impact the quality of entries into the trouble tickets. If automation can be increased, user and resource data can be copied from other databases and files without querying the users. Symptoms, however, are still necessary to fix problems. When users have more time, and are not prompted by the client contact point, they might enter more accurate data into the electronic form of the trouble ticket.

Decrease the subjectivity of the client contact point in asking and interpreting trouble symptoms. Subjectivity can be limited by providing easily understandable checklists to users on the nature of problems and probable causes. These checklists must not be technical. They can be downloaded from the client contact point to users in real time. The dialog between users and the client contact point is still there, but users are not prompted with questions. A certain level of subjectivity still remains in interpreting trouble symptoms.

Improve the quality of change requests submitted to the client contact point. Change requests must be completely filled in before being submitted. It is very important to assess the impacts of the changes; it can be done by the requester or by an instrument that accesses the configuration database for this information. Exporting this information would increase the objectivity and accuracy of judgment in approving and scheduling changes.

Decrease the inaccuracy in reporting troubles by users. Educating users on troubles, their most likely causes, and recommending actions would improve accuracy. Education is a long process, however, and because of end-user turnover, the results cannot be guaranteed. The volume of troubles could request management to implement new specific tools, such as interactive voice responses, that users do not necessarily like. These tools help to broadly categorize trivial troubles and allow more time to be spent on complex problems.

Increase the level of automation in closing trouble tickets. After successfully eliminating problems, operators tend to keep trouble tickets open or hastily close them. In either case, valuable information can get lost. For example, reason for outage is often left blank, which could be extremely important for troubleshooting similar problems in the future. Automation can import data directly from management tools and measurement instruments, offloading the operating staff from time-consuming activities and allowing them to concentrate on accurately checking the fault-closing criteria.

Increase the level of automation by correlating alarms. More sophisticated tools can increase the automation of correlating networking alarms in steps. The careful evaluation of trouble tickets helps identify related alarms. Offline expert systems can then be used to diagnose and correlate multiple and different alarms from the same family of resources, such as routers, multiplexers, modems, and bridges. The final step is the use of expert systems intelligent enough to correlate alarms across various resources.

Use more electronic work orders instead of paper-based ones. Replacing paper-based work orders with electronic forms definitely can speed up the implementation process and increase accuracy. Technically, generating and distributing work orders can be done electronically. Reviews by the planning and design group can be more easily done with the electronic form; changes, additions, and deletions are less time-consuming. Implementation and maintenance personnel can receive the work orders in time at the right place to execute the work.

Increase the level of automation for correlating measurement results for proactive problem detection. Proactive measurements are becoming very popular. In particular, the opportunities of remote monitoring (RMON) can be used to increase the automation of early warnings on threshold violations. Thresholds can be defined for almost all key performance parameters that impact the overall quality of operations. Correlating multiple parameters within a relatively narrow window requires state-of-the-art techniques, such as neural networks and expert systems.

Increase the accuracy of workload estimates for modeling. The planning and design group can improve the quality of its work if the incoming information from users is accurate. Modeling packages check the formal syntax of input, but not the feasibility of information. Inaccurate estimates cause inaccurate modeling results, with the consequence of overestimated or underestimated resource demand. Increasing the accuracy requires more education for everybody estimating workload using feasibility checks between indicators in the modeling package and then revising the workload estimates.

Increase the quality of protecting the network management systems. The combination of hardware, software, and application protection of the management system helps avoid severe security violations in the networks. In addition to the usual authorization and authentication solutions, limiting physical access to the system can help significantly. This limitation suggests the use of safe areas in network management centers. When cost-justifying additional expenditures, it is always important to remember that the most severe damages can be caused by gaining unauthorized access to the management system.

Increase the level of automation for preparing software packages to be distributed. Software distribution efficiency can substantially be improved by automating certain steps of preparing the packages at the depot prior to distribution. General rules do not work in this case; the companies must work out reasonable solutions with the vendor of the distribution software.

Network management processes. Network management processes that could be improved include change management and fault management. Each are described in the following subsections.

Change management. Conformance is not acceptable; the process flow must be improved substantially. The existing steps of the process must be decomposed, and all participants explicitly identified. Decomposition means clearly defining the paths of information export and import with other business areas. In addition, indicating time limits of management step changes would be useful. An indication of the allocation of existing and planned tools would also be useful.

Fault management. Conformance is acceptable; improvements are expected in the area of interfaces to other functions and functional groups. The existing steps of the process must further be decomposed. Decomposition means to clearly define the paths of export and import information with other business areas. In addition, indicating time limits to fault management steps would be useful. Indication of the allocation of existing and planned tools is very useful, as well. It is recommended to use multiple flowcharts, depending on the nature of the faults identified during problem detection and use separate flowcharts for hardware, software, facilities, databases, and applications.

Specific indicators. Specific indicators that can be improved upon are described in the following subsections.

Decrease the number of administrative persons in the network management organization. With a more advanced combination of documentation and administration tools, administrative persons should be cross-educated for other business areas, such as change management, provisioning, and fault tracking. Those areas also need improvement from the coverage point of view.

Improve the grade of service at the client contact point by reducing the number of poor entries. The client contact point is expected to be the single interface to clients. The perceived quality is directly correlated with the image of the whole network management organization. It is recommended to use better tools, such as automated call distributors, trouble ticketing systems, better screens for status supervision, interactive voice response, imaging support for network components, and ad hoc reporting capabilities. These steps can lead to less *poor* entries because clients will see more responsiveness and an altogether better quality of operations.

Substantially improve the proactive fault detection ratio. In this field, something must be done rapidly. It is recommended to target a combination of better tools, such as WAN and LAN monitors, that can recognize problems early. In addition, strong education is required for the staff involved in proactive monitoring. It is also recommended to clearly identify who is responsible for this activity. Improvements to reach industry average cannot be done overnight; it is expected that a time frame of 12 to 18 months is needed to observe significant changes and improvement.

Reduce the human resources expenses. Cost savings cannot be accomplished easily. Keeping salaries flat will result in more staff turnover; saving in education will result in deteriorating quality of operations. It is recommended to concentrate on stabilizing the growth of the network management staff by implementing better instruments with more capabilities of integration and automation. Network management platforms can help by reducing the number of individual managers that need to be directly managed. More powerful management applications can help reduce the time required to accomplish results. If successful, the existing team can handle more load for managing emerging technologies without significant growth in headcount.

Comparative indicators

The comparative indicators that can be improved upon are described in the following subsections.

Reduce the number of manual steps in the change management process. To make change management faster and more accurate, automation is the only answer. After the decomposition of all steps into smaller functional units, automation is easier to accomplish. It is recommended to concentrate on routine steps first. These can include checking the completeness of forms, impact analysis, feasibility, and scheduling. Products might help, but customization is absolutely necessary.

Substantially improve the proactive fault detection ratio. The weakest result is the very low proactive fault management ratio. Substantial improvement means the implementation of new instruments, such as proactive monitors in each relevant network segment, the continuous use of instruments, significant changes with skill levels, and improving the motivation of staff.

Motivation means salary additions and better job security if proactive fault detection ratios improve.

Establish more new jobs within the network management organization. Creating new job titles does not necessarily mean increasing staff and can work just the opposite. Fine-tuning job profiles and titles is helpful in assigning more specific responsibilities for organizational groups. This creation process can take a long time; the targeted time frame is around 12 to 24 months.

Increase the number of network management protocols supported. Increasing the number of protocols is not contradictory with streamlining operations. Selecting and implementing the right protocol for the right application area improves the efficiency by reducing overhead. In particular, it is recommended to use a combination of SNMPv2, CMIP, SQL, and RPC.

Improve the skill match ratio for the majority of network management functions. This improvement is the clear homework for education. Training and education of 4 to 6 weeks must be targeted, including general training, product-specific training, and on-the-job training.

Outsourcing

The companies evaluated are willing to partially outsource network management functions, starting with routine functions of operations and maintenance.

9.3   Activities to Support

The positive recommendations of the benchmarking team for the sample data are described in the following subsections.

Start innovation with a common network management database. Establishing a network management database is a very time-consuming and expensive activity. In most environments, relational and object-oriented technology are combined with each other. The database is grouped around attributes, topology information, and dynamic indicators (Figure 9.1).

Help for designing and implementing such a database can come from various sources that might, unfortunately, contradict each other. Important sources are ISO, the Internet Engineering Board, and the Network Management Forum. From the OSI perspective, a managed object is described and defined by four aspects of network management:

- Its attributes (characteristics), known at its interface (visible boundary)
- The operations that can be performed on it
- The notifications (reports) it can make
- Its behavior exhibited in response to operations performed on it

From the Internet perspective, a managed object is described in a less-abstract manner. A managed object is described by the following:

- The syntax used to model the object
- The level of access permitted to the object
- The requirements for the implementation of the object (its status)
- An unambiguous name of the object

The Network Management Forum tries to unify these definitions and provide a practically applicable list of attributes for managed objects. The list differentiates between mandatory and optional attributes. This list could become the starting point for building an integrated network management database.

Establish powerful, hierarchical filtering solutions. Absolutely no doubt exists about the necessity of reducing data collected everywhere in the network. Intelligent filtering in multiple steps would be the right answer. Figure 9.2 shows three steps of filtering.

Step One is a selection of messages to be sent to operators or to secondary storage devices. Practical experience shows that a relatively small portion of messages is needed by operations. For instance, IBM uses approximately 32 messages (out of 900) to automate systems network architecture (SNA) operations.

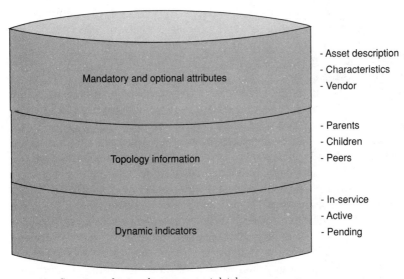

Figure 9.1 Structure of network management database

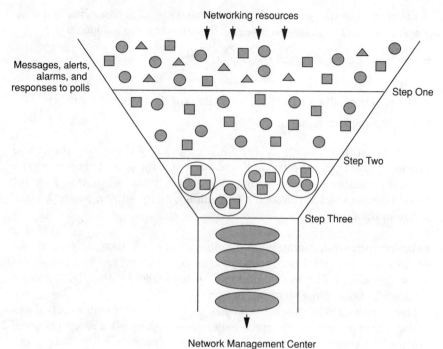

Figure 9.2 Filtering hierarchy

Step Two is to search for related messages arriving from various sources. This expert-like activity requires knowledge and skills for a large number of managed objects. Successful searching requires a careful selection of time windows, topology information about related resources, probable-cause evaluation, and severity of messages.

Step Three is the setting of priorities for related messages. For both transmission and processing of message groups, the choice of color and the choice of display for priorities are needed. High-priority messages would use expedited data transmission and other display areas on the management workstation. As a result, high-priority messages require special escalation procedures.

Looking at network management protocols, the requested intelligence of agents is very different. Common management information protocol (CMIP) could support all three steps. Simple network management protocol (SNMP) in its original version, however, could support none of them. The user should select a healthy combination of both. Both protocols allow polling and the use of traps. An optimal combination means the reduction of agent-software-storage requirements and bandwidth requirements for WAN segments. Manufacturers implementing smart agents in their networking components are on the right track in solving this problem. Figure 9.3 shows the basic principle of the operation of smart agents. Figure 9.4 displays the results accomplished by Network Control Engines from Bay Networks in reducing the polling overhead.

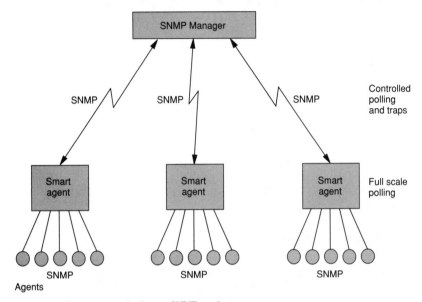

Figure 9.3 Smart agents for better SNMP performance

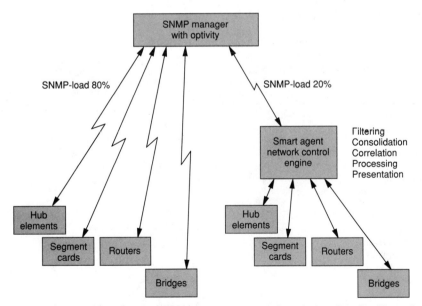

Figure 9.4 Overhead reduction by filtering using management solutions from Bay Networks

Streamline the management and measurement of indicators. Management information bases (MIBs) are the basis of collecting and storing information about various managed objects. MIBs still have special public and private areas. The private area is characteristic for the resource (managed object). The emphasis is on special features of the resource. Table 9.1 shows one recommendation about various MIBs, such as host MIB, bridge MIB, router MIB, hub MIB, and RMON MIB. These recommendations help streamline which indicators are managed and measured. These recommendations should be followed. For future applications, such as supervising desktops running under Unix, Windows, DOS, and OS/2, similar indicators are relevant. It is assumed that management platforms are available to accommodate information transferred from MIBs. Indicators outside the scope of these MIBs can still be supported, but just as part of specific measurement and performance optimization missions.

In case of desktop management, the management information format (MIF) plays the role of MIBs. In case of integrated management, MIF's must be converted into MIBs.

Find answers for strategic challenges. It is extremely important for the larger organization that managers, supervisors, and subject-matter experts clearly understand strategic issues of network management. Statements, or at least opinions, are demanded on the following items: integration, centralization, automation, standardization, databasing, and outsourcing.

Integration should be considered up to a reasonable limit. It could mean the integration of managing multiple communication forms; managing WANs, MANs, and LANs together; integrating various instruments; or integrating

TABLE 9.1 Summary of Various MIBs

Indicators	MIB II	Host MIB	Bridge MIB	Hub MIB	RMON MIB
Interface statistics	x				
IP, TCP, UDP statistics	x				
SNMP statistics	x				
Host job count		x			
Host file system		x			
Link testing			x	x	
Bridge algorithm performance			x		
Segment statistics			x	x	x
Host table				x	x
Host statistics				x	x
Historical statistics					x
Alarm thresholds					x
HostTopN					x
Traffic matrix					x
Filter					x
Capture					x

CMIP ► SNMP

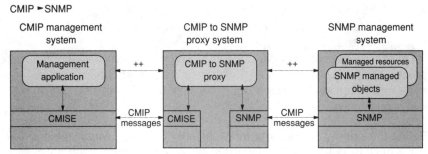

++ CMIP management system manages SNMP objects through a CMIP-to-SNMP proxy

SNMP ► CMIP

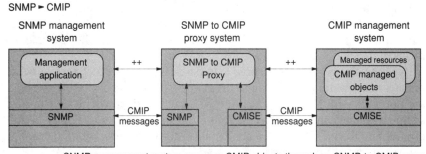

++ SNMP management system manages CMIP objects through an SNMP-to-CMIP proxy

Figure 9.5 Mutual conversions between CMIP and SNMP

the management of networks, systems, and applications. Integration is most likely not a one-step process but rather a continuous activity.

Despite downsizing and rightsizing and distributing applications to LAN-based client/server-structures, customers should keep the overall control central. Certain functions, such as data collection, data reduction, data processing, and the displaying of information, can be distributed to network management agents.

Automation and self-healing systems need attention by management. Present management products will be hopelessly overloaded unless certain alarms are answered automatically. This step does not mean "lights-out" network management centers, i.e., no personnel at all. Targeted functions are the routine tasks and the very complex tasks. Human intervention cannot be eliminated for long for those tasks of middle complexity.

The direction for standards must be addressed, and the sooner, the better. A mix of standards, consisting of proprietary and de facto, can be considered. Proxy agents help with mutual conversions. Figure 9.5 shows the mutual conversion of CMIP and SNMP using proxy agents.

Establishing and maintaining an integrated database can bring a number of benefits. It is recommended to select the attributes recommended by the

Network Management Forum. In the case of multiple relational or object-oriented databases, a directory-type service based on X.500 is expected to be implemented.

Outsourcing needs very careful consideration. Only very well organized processes should be outsourced. Regular audits and benchmarks can determine when outsourcing would offer cost benefits without risking security and service quality.

Define the priority for security management. Security management standards must be defined by top management. Tougher security measures can introduce operational overhead and can cause user dissatisfaction to some extent. The depth of security services depends on the business; military, government, and financial institutions are at the high-end, while the manufacturing industry and service institutions tend to be at the low-end in terms of investing in hardware, software, or organizational protection.

Every business, however, should do something to protect the network management system. Security features supported by the operating systems are not sufficient. Management is expected to decide in which networking segments money should be invested: legacy systems, servers, WANs, LANs, or end-user devices. Usually, a combination is the final decision. Figure 9.6 shows the investment directions.

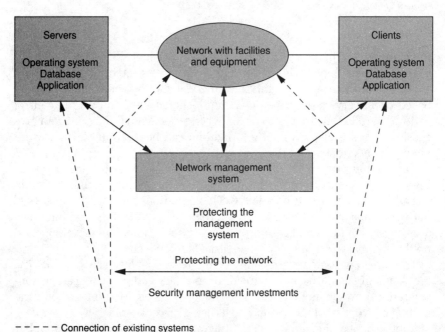

Figure 9.6 Investment directions for security management

Probability of obtaining a response time less
than a certain duration on the horizontal axis

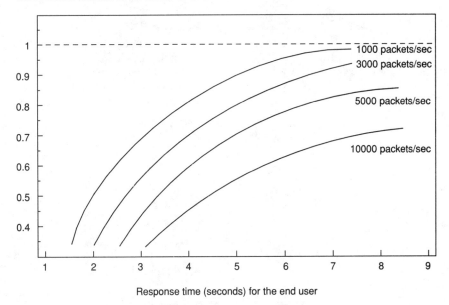

Response time (seconds) for the end user

Figure 9.7 Distribution function of response time

Quantify service quality expectations from users. Both external and internal
users are expected to quantify their service quality expectations. The indicators are manageable only when they can be measured. Recommended indicators are mean time between failures (MTBF), mean time to repair (MTBF),
response time, error rate, and throughput rates. Not only simple average values, but distribution diagrams of indicators are also important. Figure 9.7
shows the distribution function of the response time for different transaction
volumes. Segments of the network management database should be reserved
for maintaining these indicators. Also acceptable are performance databases,
such as statistical analysis system (SAS), enterprise performance data managers (EPDM), or MVS integrated control system (MICS). Figure 9.8 shows
symbolically how these databases can be populated by service indicators.

Continuously assess the value of the networks. The value of the network,
along with its network management processes, instruments, and people,
should be assessed. Communication networks are strategic assets of businesses, but networks cannot be operated without network management
processes, instruments, and human resources. It is recommended to estimate the value of the networks and their network management solutions
periodically. Comparing indicators such as networking expenditures in relation to total revenue or profit and the percentage of the communication

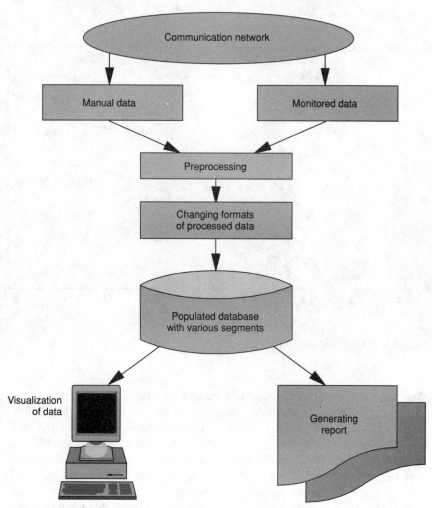

Figure 9.8 Populating performance databases

budget helps find the right position of the company in terms of investment.

Position network capacity planning and design. To optimize network management instruments, both existing and future networks must be visible to network managers. Both procedures and tools should support this activity. Once bandwidth requirements are known to management, planning and design tools can be rerun to show whether the transport of management information can be guaranteed or not. Impacts on productive traffic are not acceptable. Design alternatives include inband and outband solutions, as shown in Figure 9.9.

Select and integrate network management applications. It is recommended that multiple management applications be integrated on the same platform (Figure 9.10). Platforms are usually Unix-based with standard features such as autodiscovery, automapping, alarm support, SQL-capability, SNMPv1 and SNMPv2 support, and application programming interface support. The average enterprise deals with multiple applications that can be categorized as follows:

- Device-specific (routers, bridges, hubs, multiplexers, modems)

- Process-specific (trouble ticketing, asset management, traffic monitoring, reporting)

- Platform extensions (console emulation, presentation, eventing)

- Service-specific (ATM-management, frame-relay management)

It is recommended that these applications be integrated as deep as possible with the platform. Many alternatives exist for integration:

- Tool-bar integration to launch the tool from the main menu

- Command-line integration that offers a common area for exchanging information

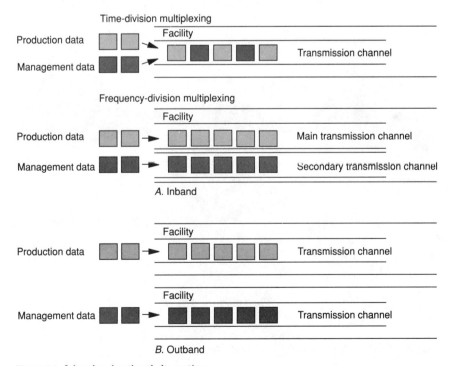

Figure 9.9 Inband and outband alternatives

- MIB integration that offers the common denominator in the MIB
- Application programming interface that offers true information exchange between platforms and applications

The average enterprise must find applications in the following fields:

- Device-dependent applications that support all principal managed objects
- Trouble ticketing that also covers change management and order processing
- Asset and cable management
- Software distribution
- Software licensing
- Traffic monitoring
- Performance reporting
- Support of relational or object-oriented databases

The manufacturers of applications offer multiple versions compatible with leading platforms from Hewlett Packard, IBM, AT&T, and SunConnect.

Start building the knowledge base with historical trouble tickets. Expert systems can assist network management by offering diagnosis, conclusions,

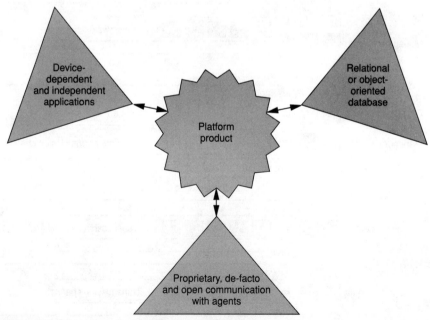

Figure 9.10 Integration of applications into platforms

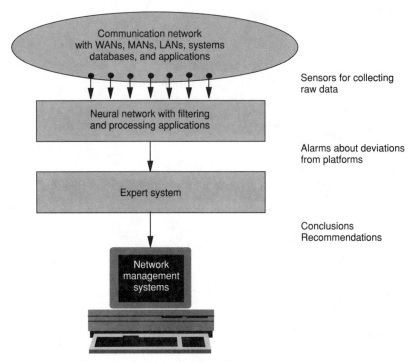

Figure 9.11 Combination of neural networks and expert systems

and recommendations for eliminating faults and problems. Future systems will most likely combine expert technology with neural networks (Figure 9.11). First implementations are recommended for offline cases that allow human operators to interfere if necessary. To reduce the time demand to program expert rules, it is recommended to use trouble tickets as source of information. Trouble tickets contain all necessary entries, such as symptoms of problems, how to solve problems, and the identification of tools and vendors involved in problem solving.

It is not necessary to relax rules. The first set of rules might be modest, but the base of rules can continuously be extended. The only prerequisite is that the expert system and its shell are powerful enough. Changing and expanding rules should be extremely user-friendly.

Apply baselining to find realistic resource utilization ranges. The central management of many local area network segments is possible technologically. Measurement probes can be installed in all segments to continuously collect data on performance indicators (Figure 9.12). These probes are not expensive and usually work together with various platforms. To avoid too much data at the platform being processed, the RMON-standard is recom-

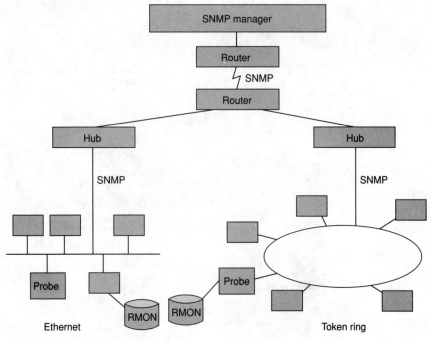

Figure 9.12   Baselining

mended for the structure of collected data. RMON is supported by both
Ethernet (Table 9.2) and token ring (Table 9.3).

TABLE 9.2 Ethernet-RMON Parameters

Group	Characteristics
Statistics	Features a table that tracks approx. 20 different characteristics of traffic on the Ethernet LAN segment, including total octets and packets, oversized packets, and errors
History	Allows a manager to establish the frequency and duration of traffic observation intervals, called "buckets." The agent can then record the characteristics of traffic according to these bucket intervals
Alarm	Permits the user to establish the criteria and thresholds that prompt the agent to issue alarms
Host	Organizes traffic statistics by each LAN node, based on time intervals set by the manager
HostTopN	Allows the user to set up ordered lists and reports based on the highest statistics generated via the host group
Matrix	Maintains two tables of traffic statistics based on pairs of communicating nodes; one is organized by sending node addresses, the other by receiving node addresses

Group	Characteristics
Filter	Allow a manager to define, by channel, particular characteristics of packets. A filter can instruct the agent, for example, to record packets with a value that indicates they contain messages from a certain network architecture (SNA, DECnet, DSA, DCA, etc.)
Packet capture	Works with the filter group to let the manager specify the memory resources to be used for recording packets that meet filter criteria
Event	Allows the manager to specify a set of parameters or conditions to be observed by the agent. Whenever these parameters or conditions occur, the agent records an event onto a log

TABLE 9.3 Token Ring RMON Parameters

Group	Characteristics
Statistics	Includes packets, octets, broadcasts, dropped packets, soft errors, and packet distribution. Statistics are at two levels: MAC for the protocol level and LLC statistics to measure traffic flow.
History	Long-term historical data for segment trend analysis. Histories include both MAC and LLC statistics.
Host	Collects information on each host discovered on the segment
HostTopN	Provides sorted statistics that allow reduction of network overhead by looking at the most active nodes
Matrix	Reports on traffic and errors between any host pair for correlating conversations on the most active nodes
Ring station	Collects general ring information and specific information for each station. General information includes ring state (normal, beacon, claim token, purge), active monitor, number of active stations. Ring station information includes a variety of error counters, station status, insertion time, and last enter/exit time.
Ring station order	Maps station MAC addresses to their order in ring
Source routing statistics	In bridges, source routing environments, information is provided on the number of frames and octets to and from the local ring, on broadcasts per route, and frame counter per hop.
Alarm	Reports changes in network characteristics based on thresholds for any or all MIBs. In this case, RMON can be used as a proactive tool.
Event	Logging of events on the basis of thresholds. Events can be used to initiate functions such as data capture or instance counts to isolate specific segments of the network.
Filter	Definitions of packet matches for selective information capture, including logical operations (AND, OR, NOT) so network events can be specified for data capture, alarms, and statistics.
Packet capture	Stores packets that match filtering specifications

Implementation examples can be found from Armon, Hewlett-Packard, Network Generals, and Axon. The platform is then responsible for further processing of the data. After a relatively short period of time, typical utilization patterns can be recognized. These patterns, combined with workload profiles, give the utilization baseline for the LAN segments under consideration. If permanent probes are not affordable, baselining can be implemented as a periodic service.

Baselining produces a statistically valid characterization of normal network behavior over an extended period rather than a specific interval, accounting for varying levels of traffic at different times. By identifying network performance through a series of statistical calculations and averages, it is possible to establish a profile, e.g., a baseline, of the LAN-segment to be used for capacity planning, problem and change management, and asset and configuration management. Figure 9.13 shows two examples using the key performance indicators of response time and line utilization.

Evaluate object-oriented technology. If networks and systems are to be managed together, storage requirements will exceed the capacity of database ca-

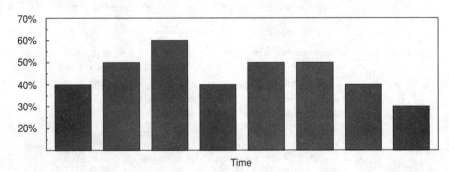

Figure 9.13 Performance report examples

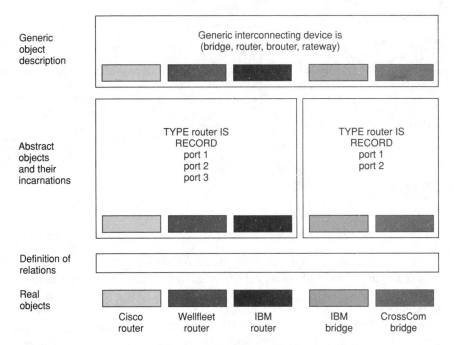

Generic object description	Generic interconnecting device is (bridge, router, brouter, rateway)

| Abstract objects and their incarnations | TYPE router IS RECORD port 1 port 2 port 3 | TYPE router IS RECORD port 1 port 2 |

Definition of relations

Real objects

Cisco router   Wellfleet router   IBM router   IBM bridge   CrossCom bridge

Figure 9.14 Saving storage requirements using object-oriented technology

pabilities of the average enterprise. Relational technology is not the only answer in these cases. *Object-oriented technology* offers a more economical use of storage capabilities. Because the code is reusable, common attributes of managed objects are stored just once. Specific attributes are maintained only for each individual object. Altogether, up to 70 percent of storage can be saved using this technology. Figure 9.14 shows an example with the object internetworking devices. It indicates that vendor-specific attributes must be specified in the last incarnation step. Relational masks are still applicable with this structure, offering database support for fault and performance management.

Share information and knowledge between processes and users. Network management functions usually use segments of data at any point in time. At other times, data are stored and usually not accessible to other functions or persons. Surveys show that over 80 percent of network-management related data are in so-called data "jails." Information and knowledge should be shared between various processes and various people. Authorization rights should govern who is entitled to use what information. Typical areas for sharing information are the following:

- Asset attributes
- Topology data

- Actual alarms
- Historical alarms
- Trouble tickets
- Historical trouble tickets
- Change requests
- Orders for networking equipment and facilities
- Workorders for provisioning
- Capacity reserves of equipment and facilities
- Backup components
- Actual resource utilization data
- Baselines for networking segments

This information should be shared within the network management organization. In certain cases, sharing might exceed the premises of the corporation.

Simplify network management processes.   Each organization uses network management processes. These processes should be subdivided into small units, such as functions, subfunctions, and applications. The allocation of instruments and people to these steps should then be precisely analyzed. Recommended guidelines are the following:

- Eliminate checker functions
- Use electronic information import and export
- Avoid media changes
- Standardize productivity tools such as word processing, E-mail, business graphics, and project management
- Analyze time for each process step
- Identify inter- and intraprocess waiting times

If implemented, process lead times and the number of persons participating in the network management processes can be substantially reduced.

Provide basic instruments for each network management business area.   Use the business model of Chapter 3 to group network management functions around the following areas:

- Client contact point
- Network operations
- Change control

- Fault tracking

- Capacity planning and design

- Finance and billing

- Fault monitoring

- Performance monitoring

- Security management

- Implementation and maintenance

- Systems administration

Basically, there are two directions for investing in instrumentation. A horizontal direction provides basic instrumentation to each business area. A vertical direction provides just basic or advanced instrumentation to a specific business area. Figure 9.15 illustrates these dimensions. It is recommended that in the first step, basic instrumentation be budgeted for each business area. In subsequent steps, investments could be divided more vertically, where advanced instrument features are required.

Use only future-proof tools. Future-proof tools are based on network management platforms. Platforms are flexible, scalable, powerful client/server structures with the capability of interfacing databases, communicating with

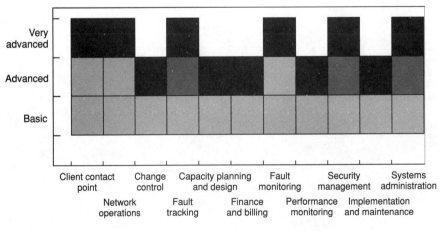

Instrumentation

First-step investments

Second-step investments

Third-step investments

Business areas of network management

Figure 9.15 Dimensions of investment

agents, and communication with various applications (Figure 9.16). Basic features of a platform are used currently, but some tools will be limited in life because of moves to other technologies and communication media. It is expected for a certain percentage of tools to become obsolete within the next 3 to 5 years. Examples of tools include the following:

- Unix-based operating system
- Autodiscovery of networking components
- Support of SNMPv1 and SNMPv2
- Offering an alarm interface
- SQL-capability to communicate with databases
- Published application programming interfaces

Some products offer even more advanced features, such as the following:

- Automapping
- Distribution of management servers
- Built-in modeling features
- Strong graphical and reporting functions

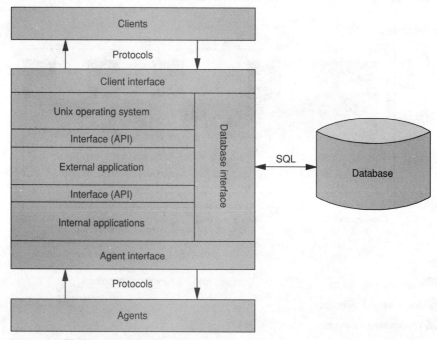

Figure 9.16 Platform structure

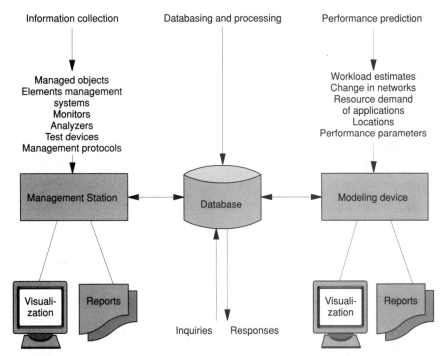

Figure 9.17 Instrumentation overview

It is recommended that a platform product or products be selected with a life expectancy of not less than five years.

Implement the right mix of instruments. No single instrument can address all needs of an average enterprise. Usually, three major groups are separated from each other (Figure 9.17):

- Information-collection instruments (Table 9.4)
- Databasing and processing instruments (Table 9.5)
- Performance prediction instruments (Table 9.6)

TABLE 9.4 Information-Collection Instruments

	Criteria				
Instruments	Complexity	Cost	Accuracy	Human resources demand	Overhead
Accounting packages	low	low	fair	low	low
Response time monitors	low	low	good	medium	low
WAN element management systems	medium	medium	good	medium	low

TABLE 9.4 Information-Collection Instruments (Continued)

Instruments	Complexity	Cost	Accuracy	Human resources demand	Overhead
MAN element management systems	high	high	good	medium	low
LAN element management systems	medium	medium	good	low	low
WAN monitors and analyzers	high	medium	high	medium	low
LAN monitors and analyzers	medium	medium	high	low	low
PBX monitors	high	high	high	high	low
Software monitors	medium	medium	medium	medium	medium
Security monitors	medium	medium	good	medium	low
Application monitors	high	variable	variable	medium	medium
Expert systems	high	high	variable	high	medium

TABLE 9.5 Databasing and Processing Instruments

Instruments	Complexity	Cost	Accuracy	Human resources demand	Overhead
Accounting packages	low	low	fair	low	low
Monitoring software	low	low	fair	low	low
Administration tools	low	variable	fair	medium	fair
Information database	medium	variable	high	high	variable
Statistical packages	medium	medium	good	high	variable
Relational databases	medium	medium	high	variable	medium
Object-oriented databases	high	high	high	high	medium
Service-level evaluation tools	high	medium	fair	high	high
Expert systems	high	high	medium	high	high

TABLE 9.6 Performance-Prediction Instruments

Instruments	Complexity	Speed	Costs	Risks	Human resources demand
Rules of thumb	low	high	low	high	low
Financial analysis tools	low	high	variable	medium	medium
Regression and correlation analysis	variable	low	variable	high	high
Time-series analysis	variable	medium	variable	high	high
Applied queuing	medium	high	medium	low	variable
Erlang-B technique for voice	low	high	low	low	low
Expert systems	high	high	high	medium	high
Simulation	high	medium	high	low	high
Benchmarks	high	medium	medium	low	high
Remote terminal emulation	high	low	high	low	high

These tables use a few criteria, such as complexity, cost, accuracy, human resource demand, and overhead, to help users preselect the right instruments.

It is recommended that the right mix of instruments be determined from the three principal groups by considering the following criteria:

- Sharing of the same database

- Integration of selected tools with management platforms

- Use of interfaces to promote information export and import among various tools

Request accurate APIs from vendors. Platform manufacturers and independent software vendors are interested in more integration between their products. The prerequisite is the publication of application programming interfaces (APIs). The present status is very unsatisfactory because of the lack of standards, different APIs, or too many API errors in the published APIs. It is recommended that accurate API-publications be requested from vendors.

Evaluate the integration depth between applications and management platforms. In integrating platforms and applications, the word *integration* can mean many things. The depth of integration is not likely to be published by the vendors. Alternatives start with menu-bar integration and end with exactly documented APIs. It is recommended that the depth of integration be checked between applications and platforms using the following checklist:

- Tool-bar integration that can launch the tool from the main menu

- Command-line integration that can offer a common area for exchanging information

- MIB integration that can offer the common denominator in the MIB

- APIs that can offer a true information exchange between platforms and applications

The ultimate goal is to work with applications using approximately the same depth, who can talk to each other by supporting information export and import. For this purpose, different protocols, such as SNMP, CMIP, SQL, and RPC, are available. The AIMS Working Group of the Network Management Forum focuses on the ad hoc communication between different applications such as alarm consolidation, event management, modeling, and database queries (Figure 9.18). It is recommended that an eye be kept on further results of this working group.

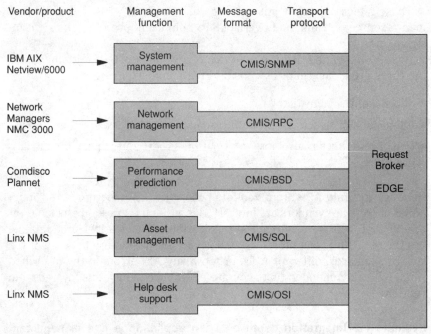

Figure 9.18 Integration by means of edges

Select platforms based on de facto or open standards. In the case of platforms, the likelihood for standardized solutions is higher than for individual
products. It is recommended that platforms be evaluated using the following criteria:

- What operating systems are in use
- What presentation services are supported
- If graphical user interfaces (GUIs) are supported
- What database management options exist
- If interprocess communication is supported
- What housekeeping functions are offered
- What management agents are supported, such as SNMPv1 or SNMPv2,
  CMIP, CMOL, CMOT, or RPC

It is recommended that the platform with the most open capabilities be
selected. The demarcation line between open and de facto is not solid, but
instead has many exceptions. For instance, agents permit the use of prox-
ies for SNMP, which demonstrates that many proprietary protocols would
be acceptable.

Discontinue with vendors not willing to migrate to standards. It is recommended that vendors be replaced when they are not willing or are unable to incrementally migrate their products to standards. The usual strategies include a retrofit of management systems and agents to SNMP capabilities, which is characteristic of WAN-management systems, including modems, multiplexers, and packet switches. The minimal requirement would be to accept the OMNIPoint recommendations of the Network Management Forum. Figure 9.19 shows the structure of this low-level but practical integration, indicating supported protocols and interfaces to applications and databases.

Test instruments thoroughly before implementation. Management platforms, monitors, and applications play a principal role in managing networks. Prior to installation, they must be thoroughly tested by the operations team using life data if possible. The tests should include functional tests, integration tests, and stress tests.

Stress tests are time-consuming and not inexpensive, but they are the only way to gather information on stresspoints and saturation levels of management platforms. Figure 9.20 displays a simple structure of a stress-testing arrangement. Particular emphasis should be on the driver; this module generates a load that emulates real managed objects. Usually, two drivers are used; one generates SNMP traps and answers SNMP polls, while the second represents the non-SNMP environment with proprietary and open protocols.

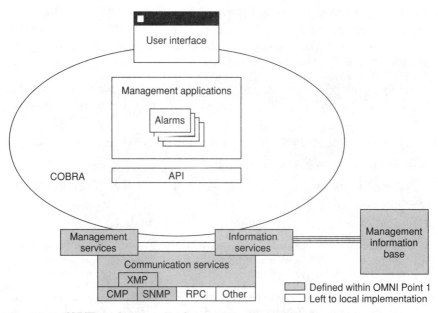

Figure 9.19 OMNIPoint for connecting heterogeneous instruments

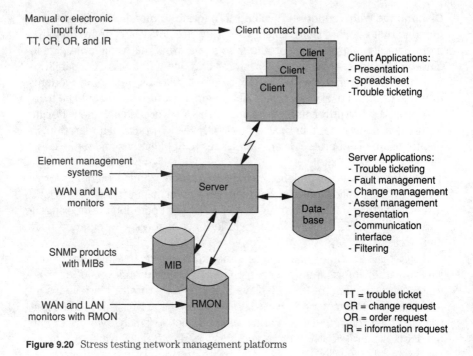

Figure 9.20 Stress testing network management platforms

Involve the future user in tool selection. Planners usually make decisions about network management related instruments. It is recommended that the actual users (clients) of the products be involved in the selection process. Typical phases of the process are the following:

- Assembly of selection criteria grouped around network management functions, performance, conformance to standards, and generic purchasing criteria

- Requests for information (RFI) sent to many vendors to preselect products

- Preselection of products and vendors

- Requests for proposals (RFP) sent to a limited number of vendors to select products

- Evaluation of proposals, including vendor presentations

- Trial installation of products from the short list (up to approximately three products) at the user premise (the future users of products must be involved in at least this phase)

- Decision using the original criteria and practical experiences with trial installation (experiences can include intangible criteria, as well)

Use proactive monitoring to recognize problems early. The use of proactive monitoring means the early recognition of hidden faults and performance bottlenecks in WANs, MANs, and LANs. The deteriorating quality of communication facilities can also be detected using monitoring. The instrumentation can be very similar to baselining; however, the emphasis is on hidden faults and not only performance indicators. If this proactive function is not continuous, the word *testing* is more commonly used for the activity. In both cases, the same indicators (lost packets, packets in error, error-free seconds, etc.) are measured and reported. Overhead could cause a problem during busy periods; measurements are very important at those times, but they can impact the productive traffic. If this activity is discontinuous, mobile probes could be used instead.

For display and reporting, platforms or individual products can be used. Integration is not the highest priority. Proactive monitoring includes systems management as well. In this case, management technology can remain the same, but other performance indicators are used. As a result, end-to-end management is supported.

Use powerful presentation techniques. Many products can compress and interpret management data. Some of the established products offer good user interfaces and fair graphics. It is recommended that the look of existing and new reports are improved. In particular, the following visualization alternatives are recommended:

- Display calendars to give daily overviews about key indicators in summary.

- Use checkerboards as an alternative to three-dimensional representations. While x and y are ordinary variables, ranges of z values should be assigned either a color or shade of gray. For each x and y value, the corresponding z value is shown by another color or shade assigned to the specific range.

- Use hierarchical diagrams to visualize device structures that are hierarchically connected. Besides trivial graphics, utilization ranges are represented by colors and shades.

Promote team spirit. Network management teams consist of many people. It is recommended that team spirit be emphasized, but individual responsibilities defined at the same time. Individual responsibility does not mean the single person is expected to execute all details. Assistance during execution can come from the inside or the outside. Typical cases for individual responsibility are the following:

- Opening, tracking, and closing trouble tickets
- Reviewing, scheduling, executing, and documenting change requests

- Reviewing, scheduling, provisioning, and testing purchase orders
- Distributing software to remote locations

Experience shows that employee motivation improves from individual assignment of responsibilities.

Compensate employees satisfactorily. Payment is a very crucial issue for employee satisfaction. Payment is still the top priority with the majority of employees. The higher the payment, the higher the quality expectation. It is recommended that employees be given industry average or slightly higher salaries so as not to lose them to the competition. Payment, however, should be linked somehow to geographical location and the skill and educational levels of the employees. It is very useful to link salary increases to continuing education.

Offer job security. Employees should see the benefits of being loyal to their employer. Besides payment, job security is the second highest priority from the perspective of employees. Job security should involve career development for both managers and subject-matter experts, planned salary increases, regular education and training, and the support of postgraduate courses and seminars.

Praise outstanding performance. Better-than-average performance must be recognized and honored. Indicators can be borrowed from technical areas to evaluate the performance of individuals, but team spirit must not be risked or violated at the cost of praising overaverage performance. Excellent managerial skills are required for this recommendation.

Train and cross-train network management staff. Continuous training is absolutely necessary. Not only instruments, but also processes, are subjects of education and training. It is estimated that one subject matter expert can operate up to four element management systems and three monitors or analyzers simultaneously. To ensure mutual substitutions during vacation, training, and sickness, cross-education is recommended. Training—even on the same platform, element management system, or monitor—should be repeated periodically. Training can be conducted by internal staff, vendors, or educational institutions. It is recommended that approximately 4 to 6 weeks of training on average be scheduled and budgeted for a person annually. This training could be a combination of in-house and off-premise courses and seminars.

Carefully rotate staff. Individual tasks can become boring and non-challenging after long periods of time. Rotating teams around job assignments can help. Consider rotation as part of continuing education. Three to six months are recommended as the duration of rotations. Less than 3 months would bring nervousness and stress into the organization; more than 6 months is risky because teams feel too comfortable with the new assignment. Additional care should be taken about the following items:

- The right sequence of jobs. The following sequence works well: client contact point –> operating –> fault monitoring –> performance monitoring –> change control –> client control point
- No changes in basic salaries.
- Written agreement by employees that they return to their original assignment.

Companies experience lowest risks when they rotate people occasionally to the performance optimization workgroup, which might not exist permanently but offers challenging tuning and optimization projects with different skills required.

Arrange meetings between end users and network management staff. Service-level agreements are the only official links between users and network management staff. Performance is judged by indicators only. To avoid tension and disagreements on the actual values of indicators, it is recommended that meetings be arranged between users and the management team. During meetings, mutual problems should be presented and explained in depth. Mutual expectations can also be addressed. Experiences show that visiting each others' working areas helps understanding even more. Meetings and visits do, however, cost working hours and require interruptions in the production at both sites.

Verify skill levels against requirements. Matrices for the necessary skill levels for each network management functional group have been presented in earlier chapters of this book. Verify skill levels against requirements periodically. The results can influence the training and educational curriculum. Gaps should openly be discussed between manager and subject-matter experts. Even subordinates can judge the managerial skills of managers. Gaps in skills can be linked to job security and salary levels.

9.4 Activities to Avoid

There are also certain activities that should be avoided when optimizing network management. These activities are discussed in the subsections below.

Avoid integration using proprietary architectures. Integration across networking architectures, network management functions, and instruments is high on the priority list of many companies. The way to integration, however, is difficult. Questions that must be answered are the following:

- With what speed is the integration expected to be accomplished?
- What will exactly be integrated?
- Who is in charge of the integration?
- What is the basis of integration?

Proprietary network management products and protocols should not be used as the basis of integration. Proprietary architectures lock companies in for many years, making them completely dependent on the manufacturer. The progress and evolution is then solely controlled by the manufacturer. Network management on its own is not a money-making business, however, which can disfavorably impact investments. Thus, companies must either implement home-grown solutions, especially for applications, or migrate to other network management architectures and products. Both choices are slow and expensive.

Avoid using a single integrator in complex networks. If integration over many instruments and functions is the ultimate target, multiple choices exist for the integration concept; single or multiple managers, management platforms, or a combination of all the above. The right choice depends on the size of the network. Experience shows that the single-manager approach can bring many risks to the company, including a single point of failure and an overloading of the integrator. It is not recommended that just one manager be used as the focal point in complex networks. Instead, it is better to implement a more distributed approach, such as distribution by managed objects, management functions, or geographical areas.

Avoid overloading operators with too many messages. All manufacturers are very proud of generating many messages about network status and performance. These messages are also routed to network management focal points. The majority of these messages, however, are redundant and do not directly contribute to solving fault and performance problems. A lack of appropriate filtering at locations and intervals hopelessly overloads operators. It is not recommended to route all types of messages to network operations. The solution is multilayered filtering by type and priority. Neural networks could deal with a very large number of messages from the network. It is recommended that approximately 90 percent of the messages be filtered out.

Avoid using multiple files and databases. The use of multiple files and databases to support network management functions is wide spread. It is recommended that the network migrate to more integrated solutions. With present solutions, the synchronization criteria cannot be solved. Because the data files and databases are from different vendors, synchronization applications require a lot of resources for programming and maintenance. Object-oriented technology, in combination with relational masks, offers a more centralized and integrated solution.

Avoid discontinuing with security risk evaluations. Distributed networks present more opportunities for security violations that can cause very serious damage to corporations. Companies must not stop analyzing security risks; it is a continuous process to evaluate threats to hardware and software. If changes are requested to hardware and software, security impacts must be immediately evaluated. Periodic security audits are recommended.

Avoid bypassing internal company accounting policies. Accounting is usually an area with tradition. Procedures have been established for relatively long periods of time, and changes and innovations are very slow. Network management needs accounting functions and instruments that could be identical, similar, or completely different from those used in other areas. It is not recommended that existing accounting solutions be bypassed just for the sake of network management—doing so would be expensive to implement and require additional skills to maintain.

Avoid using cost as the only network optimization objective. Network optimization is important for at least two reasons: improving performance and reducing costs. Besides the experience of planners and designers, modeling tools can help make the optimization process objective. Modeling tools use a number of criteria for restrictions and only one parameter as the objective function. It is not recommended that costs be used as the only objective function. Tariffs of suppliers change too quickly; it could happen that the cost-optimal solution is no longer optimal even before going into production. Budgets do, of course, play an important role, but more as a constraint rather than an objective function.

Avoid designing and developing everything in-house. The majority of network management functions can be supported by applications. There is no reason to invest in specific solutions. It is therefore not recommended to develop and implement everything in-house; rather, it is recommended that existing products and applications be selected and integrated. Integration expenses—even with professional integrators—should not be underestimated. Costs range between two to five times the cost of all hardware, software instruments, and applications purchased to support network management.

Avoid installing products that cannot support existing and evolving standards. Big competition exists between suppliers of network management products. Companies in the platform business are limited to about 10. Companies who provide management agents and applications, however, are many. The user has a real choice for the latter. It is not recommended that products be selected that cannot support existing and evolving standards.

In the area of protocols, the products should support SNMPv1, SNMPv2, CMOL, XMP, CMIP and maybe two or three of the widely used de facto proprietary standards, such as NMVT or MSU from IBM. OMNIPoint conformance would also be very useful for consolidating solutions around less protocols.

Avoid starting with sophisticated network modeling tools. The use of modeling tools in WANs, MANs, and LANs helps predict future performance and emulate saturation levels of configurations. Many products require various skill levels from users. It is not recommended to start with complex and sophisticated modeling instruments. It is better to learn the basics first, such as learning to use a few queuing equations on spreadsheets. Then follow with an easy PC-tool. After becoming unsatisfied with the efficiency and modeling limits of the tools, it is then time to migrate to more powerful products. It is very likely that the average user will use a combination of instruments that might or might not be integrated.

Avoid using too many platforms and integrators. The market of platforms, agents, and applications for network management is booming. Companies sometimes work with many element management systems that run on one or more of the leading platforms. Each system has special features, making the cross-education of network management staff even more difficult. It is not recommended that too many platforms be used because a large number of platforms require a larger operating staff.

It is recommended that the company integrate and centralize around two to three platforms that can accommodate existing element management systems and applications. These platforms are expected to communicate with each other. The communication protocols might be SNMPv2, CMOL, or a proprietary protocol. More efforts are requested in this area because the integrated, cooperative operation between multiple integrators is still very primitive and does not offer bulk data transfer, mutual alarming features, database enquiries, mutual synchronization updates, or the shared use of applications.

Avoid using too many redundant instruments. Indicators can be measured with various technologies and instruments. A certain level of redundancy is beneficial and useful for plausibility control. Redundancy does, however, cause overhead and cost increases for its purchase and maintenance. Staff is also required to operate the instruments. It is not recommended that redundant instruments be used for each group of indicators measured and supervised. On the other hand, it is strongly recommended that instruments be calibrated with tools widely used and acknowledged as accurate.

Avoid hiring an artist. Hiring network management staff is difficult. Before hiring, goals must be very clear in terms of responsibilities, job contacts, qualifying experiences, educational levels, and salary ranges. During the interviews, the goal is to fill certain positions in network management. Remembering team spirit, it is *not* recommended that individualists who do not work well in teams but instead follow their own personal goals be hired. This rule is particularly true in performance management.

Avoid building support teams from solely new hires. Network management functions are supported by various teams. The size of the teams depends on the networks, their protocols, the geographical reach, and the level of automation. Ideally, teams should show a healthy mix of experienced and novice staff. It is not recommended that teams be built just from new hires. They would be hopelessly overloaded if unexpected events occur. This fact is true for experienced staff as well, but from different operating areas.

Avoid extensive staff turnover. Keeping the team together is many times more difficult than building the management team initially. The fluctuation of staff causes financial losses to the larger organization in every case. Do not encourage excessive turnover (higher than 10 percent annually) of network management staff.

Avoid pigeon-holing network management personnel. To stabilize the size of the network management staff, cross-education is extremely important. It guarantees mutual substitutions in cases of sickness, training, and vacations. Do not discontinue cross-education. In most cases, cross-education can be done on the job using vendors and colleagues.

Avoid sharing responsibility for network management processes, functions, and instruments. A single point of responsibility in network management is extremely important to avoid finger pointing when network faults need to be detected, determined, and eliminated. Do not share responsibilities between individual persons and business groups.

Avoid hiring security officers without stringent investigations. Security management is a crucial area, and a lack of security can cause the most damage to the larger organization. Security breaches happen when procedures are not followed, instruments are not properly used, and the network management system is not sufficiently protected. All these activities are supervised and controlled by the security management staff. Thoroughly check the credentials of persons under consideration for security management staff. Even the smallest doubt about a person's background should lead to a no-hire decision.

9.5 Structure of the Final Report

It is recommended that the final report contain the following subsections and detail. The report should be submitted approximately 2 to 4 weeks after finishing the benchmarking project.

Executive summary

- Indication of how to reach and overtake industry average
- Indication of what the immediate benefits are
- Indication of what the long-term benefits are
- List of priorities for all 22 specific recommendations

Details of the benchmarking work

- Phases (all typical benchmarking phases were used in the examples)
- Interview partners (5 managers and 12 subject-matter experts were interviewed for the examples)
- Sites visited (3 sites were visited for the examples)
- Major milestones (4 weeks were spent on-site; 12 weeks total lead time)

Facts of benchmarking

- Responses from questionnaires (3 different questionnaires used in the examples)
- Measured indicators (approximately 40 indicators used in the examples)
- Interview forms used (five forms used in the examples)
- In-depth analyses of functions (19 functions out of 63 in the examples)
- Logs used
- Observation results

Descriptive indicators

Use the results from Chapter 6 for the detail here.

Comparative indicators

Use the results from Chapter 7 for the detail here.

Outsourcing statement (optional)

Use the results from Chapter 8 for the detail here.

General recommendations

Based on sections 9.3 and 9.4, a subset of recommendations can be summarized here. Examples were provided in those sections.

Project plan

Project plans usually take the form shown in Table 9.7.

The final report can contain further details in the appendices. Many clients are very interested in receiving the data prior to it being processed by the benchmarking team.

9.6 Presentation of Results

The presentation should follow the highlights of the benchmarking report, concentrating on the following aspects:

- Major milestones of project
- Facts describing project and company
- In-depth discussion of descriptive indicators
- In-depth discussion of outsourcing alternatives, if any
- In-depth discussion of comparative indicators
- Summary of generic recommendations
- Summary of specific recommendations
- Project plan for performance improvements
- Questions and answers

After the presentation, the benchmarking team might be requested to slightly modify the final report. The distribution of the benchmarking report and the invitation to the final presentation is the responsibility of the project manager of the company and not the benchmarking team. The presentation

TABLE 9.7 Basic Structure of Project Plan

Activity	Priority	Schedule	Equipment demand	Human resources demand
Increase the proactive monitoring ratio for fault management	high	Implementation: IV. Quarterly 1995	Additional monitors $150,000	2 FTE (full-time equivalent) $300,000 p.a.

could become highly political, and the benchmarking team must not involve itself in company politics.

9.7 Summary

Preparing, submitting, and presenting the results is the final phase of the benchmarking process. These activities must not be underestimated. Poor presentation of the valuable results can destroy all the project work, so carefully select the technical depth of the report and presentation and the language.

It is sometimes advantageous to sponsor two presentations: one to the technical staff and a second one to the managers. During the second presentation, the experiences of the first presentation can fully be used.

Don't be surprised if the presentation results are heavily questioned by the attendees. When gaps are described—and there are always gaps—somebody feels responsible for them. These persons want to defend themselves, using excuses and blaming others. The benchmarking team very frequently must serve as the mediator during those nontechnical discussions. The best tactic is to lead the discussion to the redistribution of responsibilities and to the justification of the project plan to eliminate the gaps.

10

Trends

Auditing and benchmarking have been successfully implemented for many years for financial processes across industries. Comparing single telecommunication systems using independent benchmarks is widely used as part of the purchase process, but auditing and benchmarking for effective network management is new.

10.1 Growing Popularity of Network Management Benchmarks

The auditing and benchmarking of network management are becoming increasingly popular for the following reasons:

- The importance of communications networks to businesses is constantly growing: this importance can be highlighted by the facts that follow. Most companies understand these needs and deal successfully with planning, designing, implementing, and operating communication networks:

  ~Users and applications request any-to-any connections
  ~New applications require new communication technologies, such as video, frame relay, asynchronous transfer mode (ATM), broadband ISDN (B-ISDN), Sonet, etc.
  ~Client/server systems cannot be implemented without networks to connect them
  ~Distribution of databases cannot be accomplished without reliable networking

- The importance of network management for these communication networks is constantly growing. This importance can be highlighted by the following facts:

  ~Complex networking structures of WANs, MANs, and LANs make them no longer practically manageable.

  ~Investments in communication technologies are significant; economic use of these technologies is a top priority for the board of directors.

  ~Downtime of communication networks causes substantial financial losses to companies; powerful applications help reduce these losses.

  ~Network management systems help the early recognition of bottlenecks, and help make the right capacity upgrade decisions.

  ~Vendor products without management agents cannot be sold anymore; companies are expected to select powerful platforms that can accommodate and coordinate a number of agents.

- The expenditures of network management are growing rapidly targeting 15 to 20 percent of the communication budget. Thus, the following must be decided:

  ~How to invest this amount of money to guarantee the highest effectiveness

  ~How to justify the investment by showing the consequence of gaps in network management processes, instruments, and human resources

- Outsourcing decisions require an accurate status of present operations; the input can easily come from the benchmarking analysis. Benchmarking also helps the following:

  ~Identifying gaps; the gaps can be filled using insourcing or outsourcing

  ~Using the recommendations for performance improvements as a criteria list to evaluate the outsourcers

  ~Identifying those network management functions where outsourcing can substantially improve the economy of scale

- Comparison with best practices helps identify gaps for immediate improvements:

  ~The ideal model gives the ultimate target for performance improvements.

  ~The practical model gives more realistic targets on the basis of the average of best practices.

10.2 Justification of Benchmarks

To quantify and justify expenditures for benchmarks, companies must schedule 4 to 6 weeks lead time. The benchmarking team is usually not large; usually two to four persons are involved. The team works very closely with managers and subject-matter experts in the company. The final price of the benchmarking service depends on the country. In most cases, external resources are used for auditing and benchmarking to guarantee the unbiased view of the benchmarking results. Some consulting companies accommodate such departments; at this moment, the benchmarking teams concentrate more on rightsizing and downsizing by client/server structures rather than on networking and network management.

Practical work is pursued in network management by the following companies: AT&T Business Communication Services and Global Information Services, Compass Inc., Real Decisions (part of the Gartner Group), Nolan Consulting, Andersen Consulting, and Lynx Technologies, Inc.

Unfortunately, benchmarking results are rarely published. In the future, I expect considerably more opportunities for benchmarking teams for the following reasons:

- High network management expenditures; the role of an unbiased company is important in reviewing the direction of investments. The selection process of technologies, services, and providers is becoming more difficult because many platforms and applications are available; companies must be sure they have made the right choices.

- Network management benchmarks can be combined with quality management; ISO-9000 and TQM evaluations can embed network management benchmarks in the future. More companies are interested in participating in joint benchmarking using the same database filled with indicators. These databases can be segmented by industries.

- Internal company benchmarks will become popular because headquarters might want to standardize on processes, procedures, instruments, and human skill levels by benchmarking the business units.

10.3 Technology Changes

In the future, certain aspects of networking technology will happen more quickly because of a higher level of automation in network management. Special bandwidth provisioning, which must now be scheduled days in advance, will happen in minutes. Service restoration procedures will be initiated by expert-like failure analysis mechanisms and conducted automatically. New service provisioning will take place as the direct result of keystroke entry by a technical representative or user using a workstation-based service-creation tool. Users will have near real-time control over the dynamic reconfiguration of services. New services will be developed and implemented more rapidly than they are today by the use of preexisting systems integration.

Groups of service providers will form partnerships (e.g., AT&T and Unisource in WordSource, U.S. Sprint and Eunetcom, British Telecom and MCI) and quickly deploy automated services that require the sharing of management information across multiple post, telegraph, and telephone

(PTT) administrations. Voice and fax services are now becoming commodities offered to the public; these services are usually separated from others.

A greater number of element management systems will be purchased directly from network element vendors or systems integrators that specialize in a specific area of management; these systems will be deployed and rapidly integrated into existing management hierarchies because of interoperable interfaces. In addition, network element vendors, system integrators, and independent software vendors will offer a variety of network management applications to increase the power of existing management platforms.

Interoperability, integration, and automation will be major issues in network management for the remainder of this decade. A practical method of achieving administration-wide and industry-wide interoperability of management systems has emerged, which constitutes a paradigm shift in networking and network management technology that cannot be ignored. Vendors and service providers who are not supporting interoperability will be placed at a competitive disadvantage. As users, systems integrators, and service providers begin to set goals for the integration of management platforms and applications, they will begin to request standards-based interoperability interfaces from vendors of specialized systems.

10.4 Standardization

Isolated management, still widely used in private businesses for efficiency, but which focuses only on a narrow range of equipment issues and cannot offer interoperability, will be considered obsolete. Common examples are fault monitoring systems, test systems, event and alarm correlation applications, performance monitoring and archiving systems, signaling network managers, transmission system managers, and vendor-supplied element management systems. The industry will be best served if the competing vendors form working groups within the Network Management Forum, for example, and develop common managed objects for future use by their own developers. Vendors can apply this open approach demanded by commercial users and service providers in the form of interoperable interfaces. They can distinguish themselves in the market by the relative performance and feature sets of their standards-based offerings. Network management benchmarks help accelerate this standardization process.

10.5 Summary

Benchmarking can quantify the performance of network management functions, instruments, protocols in use, and human resources responsible for the functions, instruments, and protocols. Benchmarking for effective network management is not free of charge. External and internal resources create expenses, but benefits and middle and long-range savings outweigh these expenses with the usual result of very short payback periods.

Preliminary Questionnaire

1 Company Profile

1.1 What business is the company in?

- Finance
- Manufacturing
- Transportation
- Government
- Retail
- Telecommunications supplier

1.2 What are the major locations (sites) of the company with communication needs?

- Major divisions and business units
- How businesses vary within each division
- Spin-off divisions
- Plans to acquire new businesses

1.3 What communication needs exist today and in the future?

- Voice communication needs
- Data communication needs

- Video communication needs
- Image communication needs
- E-mail communication needs

1.4 What principal applications exist today?

1.5 What are the communication needs of these applications?

1.6 What applications are supported by which locations (sites)?

1.7 What is the estimated traffic volume between locations (sites) for each communications form?

2 Existing Networks

2.1 Data networks

2.1.1 Is the network hierarchical—point-to-point—or a combination of backbone and access networks? See Figure A.1 for an example of hierarchical networks.

2.1.2 Is the network peer-to-peer? See Figure A.2 for an example of peer-to-peer networks.

2.1.3 Is the network a combination of both logical alternatives sharing a common physical network?

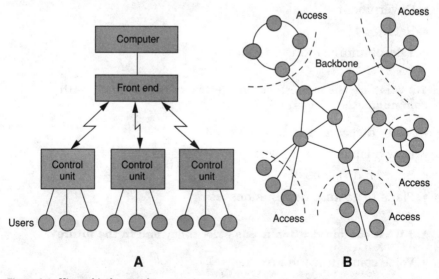

A   B

Figure A.1 Hierarchical networks

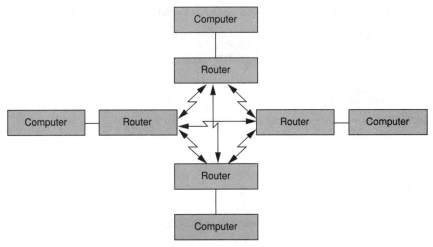

Figure A.2 Peer-to-peer networks

2.1.4 How many domestic and international networks are in operation?

2.1.5 What architectures are supported in the following networks?

- Wide area network: SNA, DNA, DCA, TRANSDATA, DSA
- Metropolitan area network: FDDI, DQDB
- Local area network: Ethernet, token ring, token bus

2.1.6 What protocols are in operation?

- Routed protocols: TCP/IP, IPX/SPX, Decnet, XNS, AppleTalk, others
- Nonrouted protocols: NetBios, LAT, Bisynch, SNA/SDLC, others

2.1.7 What principal operating systems are in use?

- Mainframes: MVS, VM, Unix, BS2000, others
- Servers: Unix, Windows, Windows-NT, OS/2, DOS
- Clients: Unix, Windows, Windows-NT, OS/2, DOS
- Network operating systems: Novell, Banyan, LAN Manager,
- Windows-NT

2.2 Voice networks
2.2.1 Architecture and structure of voice network

2.2.2 How many domestic and international networks are in operation?

2.2.3 Is the bandwidth shared with data networks?

2.3 Other networks

2.3.1 What other networks are in operation?

- Video networks
- Cellular networks
- Satellite networks

2.3.2 What applications are supported by these networks?

2.3.3 If the company has more than one network:

- Are multiple voice networks linked?
- Are multiple data networks linked?
- Are voice and data networks integrated?
- Are multiple mail networks linked?
- Are multiple video networks linked?
- Are all the above linked to each other?

2.4 What backup strategies and components are in use?

- At the physical level?
- At the logical level?
- At the data backup by voice channels?
- At the public networks?
- At the ISDN facilities?

2.5 Provide data on suppliers for network equipment, facilities, and services.

3. Transmission Facilities and Data Rates for Networks

3.1 Domestic lines

3.1.1 Backbone networks

- Analog circuits
- Digital circuits
- Switched digital

3.1.2 Access networks

- Analog circuits
- Digital circuits
- Switched digital

3.1.3 Peer-to-peer networks

- Analog circuits
- Digital circuits
- Switched digital

3.2 International networks

3.2.1 Backbone networks

- Analog circuits
- Digital circuits
- Switched digital

3.2.2 Access networks

- Analog circuits
- Digital circuits
- Switched digital

3.2.3 Peer-to-peer networks

- Analog circuits
- Digital circuits
- Switched digital

3.3 What value-added services are in use?

- Packet switching
- Frame relay
- Electronic mail

- LAN-to-LAN bridging service
- ISDN
- Managed data networks

3.4 Who are the providers of the value-added services?

- AT&T WorldPartner
- Eunetcom
- Unisource
- IBM-Advantis
- Telenet
- Tymnet
- DataPac
- Infonet
- Geis
- Concert
- Sita
- EDS
- Others

4 Networking Equipment

4.1 How many WANs are in use and who is the vendor for the following?

- Multiplexers
- Modems
- Packet switches
- Front-end processors
- Matrix switches
- Packet assembly/disassembly
- Satellite nodes (VSAT)
- Frame-relay nodes
- ATM nodes
- Concentrators
- Protocol converters

4.2 How many LANs are in use and who is the vendor for the following?

- Ethernet
- Token ring
- Token bus
- Arcnet
- Routers
- Bridges
- Extenders
- Gateways

5 Personnel

5.1 Is the organization centralized or decentralized?

5.2 What functions are supported?

Provide job descriptions and, if possible, a diagram of the network management organization for the following jobs. Also list how many people are in each type of position.

- *Management*: administration and management of network personnel
- *Planning and design*: planning for periods of 2 to 5 years in the future and analyzing new technology; designing and configuring the network
- *Operations*: monitoring and supervising of network status and performance; collecting and analyzing performance and availability statistics; diagnostics, tests, recovery, and repair
- *Client contact point*: interfacing with users regarding problem determination; providing first-tier support
- *Administration*: handling accounting, billing, order processing, inventory control, maintenance, and applications support

6 Costs

Provide estimates for the number of managed objects in the network related to total size of the networks. Provide both fixed costs and operating costs.

6.1 Transmission and communication facilities

6.1.1 Domestic lines

- Backbone networks
- Access networks
- Peer-to-peer networks

6.1.2 International networks

- Backbone networks
- Access networks
- Peer-to-peer networks

6.1.3 Value-added services in use

- Packet switching
- Frame relay
- Electronic mail
- LAN-to-LAN bridging service
- Managed data networks

6.1.4 Providers of value-added services

- AT&T WorldPartner
- Eunetcom
- Unisource
- IBM-Advantis
- Telenet
- Tymnet
- DataPac
- Infonet
- Geis
- Concert
- Sita
- EDS
- Others

6.2 Communication hardware

Provide fixed costs, depreciation, operating costs, and maintenance costs for the following.

6.2.1 Wide area networks

- Multiplexers
- Modems
- Packet switches
- Front-end processors
- Matrix switches
- Packet assembly/disassembly
- Satellite nodes (VSAT)
- Frame-relay nodes
- ATM nodes
- Concentrators
- Protocol converters
- Network management
- Others

6.2.2 Local area networks

- Ethernet
- Token ring
- Token bus
- Arcnet
- Routers
- Bridges
- Extenders
- Gateways
- Network management
- Others

6.3 Communication software

Provide fixed costs, depreciation, operating costs, and maintenance costs for the following:

- Telecommunication access methods
- Routing software
- Operating systems
- Inventory control
- Databases
- Network management
- Others

6.4 Personnel

Provide salaries, training costs, travel costs, benefit costs, and overhead/occupancy costs for the following:

- Management
- Planning and design
- Client contact point
- Operations
- Administration
- External staff

6.5 Infrastructure

Provide fixed costs, operating costs (value-added-tax, taxes, interests, etc.), and maintenance for the following:

- Cabling Hubs
- Power supply
- Security surveillance
- Backup components

7 Network Management

7.1 What network management functions are supported?
7.2 What network management instruments are in use?

- Integrators (e.g., NetView)
- Element Management Systems for WANs and LANs (e.g., modems, multiplexers, bridges, routers, hubs, PBXs, others)

- Management platforms (e.g., OpenView, SunNet Manage, NetView for AIX, OneVision, others)
- NetView/6000, StarSentry, others)
- WAN monitors
- WAN analyzers
- LAN monitors
- LAN analyzers
- Security management systems
- Network design tools
- Network optimization tools
- Network management databases
- Reporting software
- Client contact point instruments

8 Outsourcing

- Are any parts of network management already outsourced?
- What is the willingness to outsource?
- Which functions could be outsourced?
- Who could be the outsourcer?
- What management instruments would be outsourced?
- What percentage of human resources would be outsourced?

On-Site Interview Questionnaire

Part I. Investments and Organization

1. What is the general network management investment philosophy of the firm for the short term, medium term, and long term?

2. What types of criteria are used to approve network management investments for payback, present value, and return on investment?

3. What are the detailed organizational diagrams at the corporate, business unit, and network management level?

4. Are international and national voice, data, image, video and E-mail networks managed by the same group?

5. What is the mission of the information services group?

6. Who is responsible for overall network management?

7. Who is responsible for network management related business decisions?

8. Who is responsible for strategic and tactical planning?

9. Who is responsible for operations?

10. What is the mission of the network management group?

11. What are operational goals of the network management group?

12. How many personnel are directly involved in network management operations: 1–6, 7–15, 16–50, 51–99, 100–249, or more than 250?

13. What type of general skills do you look for in the network management group?

14. What are the basic network management functions performed by the network management group?

15. What instruments are used by the network management group?

16. Who are responsible for the following functions, a specific business unit or the corporate information systems (IS) group?

   - Planning and design
   - Operations
   - Capacity planning
   - Setting standards
   - Strategic planning
   - Administration
   - Quality assurance

17. How many people outside the IS organization are involved in decision making?

18. Is the network management center considered a cost or profit center?

19. What is the annual budget including facilities and equipment for telecommunications: $500K–$750K, $750K–$1M,$1M–$2M, $2M–$5M, $5–$20M, or more than $50M?

20. What percentage of this budget is dedicated to network management: less than 1%, 1–5%, 5–10%, 10–20%, or more than 20%?

21. What are application priorities for users in backup and disaster cases?

22. What is the estimated lost revenue when LANs, WANs, and MANs are out of order, meaning that specific application services cannot be supported for users?

23. Have you ever built a business case for network management investments?

Part II. Network Management Functions and Problems

1. Define network management as it relates to your firm.

2. A generally accepted definition of network management is the deploying and coordinating of resources to plan, operate, ad-

minister, analyze, evaluate, design, and expand communication networks to meet service-level requirements. Do you agree with this definition? If not, how would you modify it?

3. How does network management relate in your company to service and systems management?

4. Listed below are the principal network management functions. State your agreement or disagreement with the functions and modify them if necessary.

- Client contact point

  ~Receiving problem reports
  ~Handling calls
  ~Handling enquiries
  ~Receiving change requests
  ~Handling orders
  ~Making service requests
  ~Opening and referring trouble tickets
  ~Closing trouble tickets

- Operations support

  ~Determining problems using trouble tickets
  ~Diagnosing problems
  ~Taking corrective actions
  ~Repairing and replacing
  ~Referencing to third parties
  ~Backing up and reconfiguring
  ~Recovering
  ~Logging events and performing corrective actions

- Fault tracking

  ~Tracking manually reported or monitored faults
  ~Tracking the progress and escalation of problems if necessary
  ~Distributing information (including reports indicating fault avoidance)
  ~Referring problems (with predictive capabilities)

- Change control

  ~Managing, processing, and tracking service orders
  ~Routing service orders
  ~Supervising the handling of changes

- Planning and design

  ~Analyzing needs
  ~Projecting application load

~Sizing resources
~Authorizing and tracking changes
~Raising purchase orders
~Producing implementation plans
~Establishing company standards
~Maintaining quality assurance

- Finance and billing

~Asset management
~Costing services
~Client billing
~Usage and outage collection
~Rebate calculation for clients
~Bill verifications
~Software license control

- Implementation and maintenance

~Implementing change requests and work orders
~Maintaining resources
~Inspection
~Maintaining configuration database
~Provisioning

- Fault monitoring

~Monitoring systems and network for proactive problem detection
~Opening additional trouble tickets
~Referring trouble tickets

- Performance monitoring

~Monitoring systems and networks performance
~Monitoring SLAs
~Monitoring third-party and vendor performance
~Optimizing, modeling, and tuning
~Reporting usage statistics and trends

- Security management

~Threat analysis
~Administration
~Detection
~Recovery
~Protection of the management systems

- Systems administration

~Software version control
~Software distribution

~Systems management
~Administration of user-definable tables
~Local and remote configuration of resources
~Management of names and addresses
~Applications management

5. Rank the above functions from 1 to 11 as to their importance in your firm.
6. What percentage of your communications budget is dedicated to network management: 0%–5%, 6%–10%, 11%–20%, or 20% or more?
7. What are the most important problems managing your networks today? List by priority and group them around processes, instruments, and people.
8. What are the most important problems managing your networks in the future? List by priorities and group them around processes, instruments, and people.
9. What are the domestic network management requirements?
10. What are the global network management requirements?
11. Between which countries would you like to have network management capabilities?
12. Between which countries do you currently have network management capabilities?

Part III. Network Management Implementation

1. How are your networks currently managed?
   - Centrally
   - Decentrally
   - At a network element level
   - At the equipment level
   - Physical layers only
   - Logical layers only
   - Not managed at all

2. Do you use a client contact point?
3. How is your client contact point staffed?
4. What instruments support the client contact point?
5. Which group within your company is responsible for each of the 11 business areas of network management?
   - Client control point
   - Operations support
   - Fault tracking
   - Implementation and maintenance

- Fault monitoring
- Performance monitoring
- Security management
- Systems administration
- Change control
- Planning and design
- Finance and billing

6. How many "full-time equivalents" (one person, full time, for a complete year, excluding training, vacation, and sickness) are involved in network management: 0–5, 6–15, 16–30, 31–50, 51–100, 101–200, or more than 200?

7. What is the annual budget for network management, including hardware, software, and staff?

8. What are the salary ranges for network management personnel in each of the 11 business areas?

9. What type of network management training is available to your network management personnel in each of the 11 business areas?

10. Who provides network management training?

- Equipment vendor
- Network services supplier
- Consultant
- Educational firm
- In-house
- Other (please specify)

11. What is the turnover rate of your network management staff: 0–5%, 6%–15%, 16%–30%, 31%–50%, or higher?

12. Describe your current approach to network management:

- Manager of managers using one focal point
- Manager of managers using multiple focal points
- Applications-based using platform
- Combination

13. What specific network management tools are used by your company today?

- Integrators
- Element management systems
- Monitors
- Administration tools

- Network design and planning tools
- Others

14. What specific network management tools are planned for the future?

   - Integrators
   - Element management systems
   - Monitors
   - Administration tools
   - Network design and planning tools
   - Others

15. Which of the tools used in network management are most effective?

16. Which of the tools used in network management are least effective?

17. Are outside parties engaged to perform any network management functions?

18. If yes, which business areas are supported by which third parties?

19. Which third parties would you prefer if you outsource network management functions?

   - Systems integrators
   - Network services providers
   - Systems manufacturers
   - Consulting companies

20. What criteria would you use to select outsourcers?

Part IV. Network Management Direction

1. How important is managing a multiple vendor environment?

2. Are you migrating toward an open systems interconnection (OSI) environment today?

3. Will OSI-based management be important in the future in 2–3 years, 4–6 years, or more than 6 years?

4. How important is the NetView product family from IBM for managing networks today?

5. How important will the NetView product family from IBM be for managing networks in the future?

6. How important is Solve:Automation from Sterling Software for managing networks today?

7. How important will Solve:Automation be from Sterling Software for managing networks in the future?

8. How important is the simple network management protocol (SNMP) in managing networks today and tomorrow?

9. How important is OMNIPoint in the network management product structure today and tomorrow?

10. What are the preferences for attributes of network management platforms today and in the future?

11. What network management platforms are preferred?

- OneVision
- OpenView
- NetView for AIX
- SunNet Manager
- DiMONS 3G
- Spectrum
- NetExpert
- MAXM

12. What network management applications are important?

- Device-specific
- Process-specific
- Service-specific
- Platform extensions
- Systems management
- Traffic monitoring and analysis
- Application development

13. What level of application integration is targeted?

- Application launching
- Command-line interface
- Event-level integration
- Database or MIB-level integration
- Standardized API

14. Who is responsible for integration?

- Client
- Manufacturers
- Independent software vendors
- Systems integrators

Questionnaire for Network Management Functions

Generic questions need to be asked during the interviewing process. The checklist for each function consists of these generic questions, as well as a number of specific questions, which is presented in this appendix.

1. Scope of function
2. Time limitations of function
3. Frequency of executing function
4. Person in charge of execution, support, and advice for function
5. Number of persons involved in execution, support, and advice
6. Documentation forms used to support function
7. Instruments used to support function
8. Level of automation with function
9. Decision-making criteria used
10. Priorities used within function for individual tasks
11. Source of information for function
12. Quality of information received for function
13. Destination of sending information from function
14. Quality control of outgoing information

Specific questions are added to the generic ones if necessary.

Client Contact Point

Receiving problem reports

Generic questions

1. Scope of function (what problems are expected to be received by client contact point)
2. Time limitations of function
3. Frequency of executing function driven by callers and periodic polling of certain regular callers
4. Person in charge of execution, support, and advice for function
5. Number of persons involved in execution, support, and advice
6. Documentation forms used to support function
7. Instruments used to support function:
   - Registration
   - Recording
   - Interactive voice response
8. Level of automation with function
9. Decision-making criteria used
10. Priorities used within function for individual task
11. Source of information for function
12. Quality of information received for function:
    - Symptoms identified by callers
    - E-mail messages from users
13. Destination of sending information from function:
    - Operational support
    - Outside vendors
    - Third-party consultants
14. Quality control of outgoing information

Specific questions   None

Handling calls

Generic questions

1. Scope of function
2. Time limitations of function

3. Frequency of executing function:
   - Driven by callers
   - Polling critical users

4. Person in charge of execution, support, and advice for function

5. Number of persons involved in execution, support, and advice

6. Documentation forms used to support function

7. Instruments used to support function:
   - Registration
   - Answering machines
   - Recorders or voice mail

8. Level of automation with function

9. Decision-making criteria used:
   - Taking or not taking call
   - Which call to take first

10. Priorities used within function for individual tasks

11. Source of information for function

12. Quality of information received for function

13. Destination of sending information from function

14. Quality control of outgoing information

Specific questions   None

Handling enquiries
Generic questions

1. Scope of function

2. Time limitations of function

3. Frequency of executing function:
   - Driven by caller
   - Polling critical users

4. Person in charge of execution, support, and advice for function

5. Number of persons involved in execution, support, and advice

6. Documentation forms used to support function

7. Instruments used to support this function:
   - Registration
   - Answering machines
   - Recorders or voice mail

8. Level of automation with function
9. Decision-making criteria used:
   - Taking or not taking call
   - Which call to take first
10. Priorities used within function for individual tasks
11. Source of information for function
12. Quality of information received for function and clearness of enquiry
13. Destination of sending information from function
14. Quality control of outgoing information

Specific questions   None

Receiving change requests
Generic questions

1. Scope of function (what changes are considered)
2. Time limitations of function
3. Frequency of executing function:
   - Driven by requesters
   - Daily or weekly schedules
4. Person in charge of execution, support, and advice for function
5. Number of persons involved in execution, support, and advice
6. Documentation forms used to support function
7. Instruments used to support function
8. Level of automation with function
9. Decision-making criteria used
10. Priorities used within function for individual tasks
11. Source of information for function
12. Quality of information received for function:
    - Completeness of forms
    - Legibility if handwritten
13. Destination of sending information from function
14. Quality control of outgoing information

Specific questions

1. Persons authorized to request changes
2. Persons authorized to modify changes

Handling orders

Generic questions

1. Scope of function
2. Time limitations of function
3. Frequency of executing function:
   - Driven by requesters
   - Periodic collection of orders (daily, weekly)
4. Person in charge of execution, support, and advice for function
5. Number of persons involved in execution, support, and advice
6. Documentation forms used to support function
7. Instruments used to support function:
   - Voice mail
   - E-mail
   - EDI
   - Letter (fax, telex)
8. Level of automation with function
9. Decision-making criteria used
10. Priorities used within function for individual tasks
11. Source of information for function
12. Quality of information received for function
13. Destination of sending information from function
14. Quality control of outgoing information

Specific questions

1. Number of orders generated within a given period of time

Informing clients

Generic questions

1. Scope of function (what information is going to be distributed to clients, such as status, performance, etc.)
2. Time limitations of function
3. Frequency of executing function
4. Person in charge of execution, support, and advice for function
5. Number of persons involved in execution, support, and advice
6. Documentation forms used to support function
7. Instruments used to support function
8. Level of automation with function
9. Decision-making criteria used
10. Priorities used within function for individual tasks
11. Source of information for function
12. Quality of information received for function
13. Destination of sending information from function
14. Quality control of outgoing information (level of detail of information and format)

Specific questions

1. What communication forms are used to distribute information

Making service requests

Generic questions

1. Scope of function
2. Time limitations of function
3. Frequency of executing function
4. Person in charge of execution, support, and advice for function
5. Number of persons involved in execution, support, and advice
6. Documentation forms used to support function
7. Instruments used to support function
8. Level of automation with function
9. Decision-making criteria used

10. Priorities used within function for individual tasks
11. Source of information for function
12. Quality of information received for function
13. Destination of sending information from function
14. Quality control of outgoing information

Specific questions

1. Communication forms used to make service requests

Opening and referring trouble tickets
Generic questions

1. Scope of function (what troubles are included and excluded)
2. Time limitations of function
3. Frequency of executing function
4. Person in charge of execution, support, and advice for function.
5. Number of persons involved in execution, support, and advice.
6. Documentation forms used to support function
7. Instruments used to support function
8. Level of automation with function:
   - What information is copied automatically into ticket from alarm
   - What information must be entered manually
9. Decision-making criteria used
10. Priorities used within function for individual tasks
11. Source of information for function
12. Quality of information received for function
13. Destination of sending information from function
14. Quality control of outgoing information

Specific questions

1. Criteria for opening trouble tickets
2. How much of function can be distributed to multiple locations

Closing trouble tickets

Generic questions

1. Scope of function (criteria of closing, such as testing, and confirmation of functionality)
2. Time limitations of function
3. Frequency of executing function
4. Person in charge of execution, support, and advice for function
5. Number of persons involved in execution, support, and advice
6. Documentation forms used to support function
7. Instruments used to support function
8. Level of automation with function:
   - Automated closing procedure
   - Manual closing procedure
   - What information about resolution process copied into trouble ticket
9. Decision-making criteria used
10. Priorities used within function for individual tasks
11. Source of information for function (who gives the information to close)
12. Quality of information received for function
13. Destination of sending information from function (who receives closed trouble tickets)
14. Quality control of outgoing information

Specific questions   None

Operations Support

Determining problems using trouble tickets

Generic questions

1. Scope of function (what managed objects are subject of problem determination)
2. Time limitations of function
3. Frequency of executing function
4. Person in charge of execution, support, and advice for function

5. Number of persons involved in execution, support, and advice

6. Documentation forms used to support function

7. Instruments used to support function

8. Level of automation with function

9. Decision-making criteria used

10. Priorities used within function for individual tasks:
    - Arriving trouble tickets
    - Referred trouble tickets

11. Source of information for function:
    - What components can deliver status data
    - What components must be polled for status data

12. Quality of information received for function

13. Destination of sending information from function

14. Quality control of outgoing information

Specific questions None

Diagnosing problems
Generic questions

1. Scope of function

2. Time limitations of function (before referral or escalation)

3. Frequency of executing function

4. Person in charge of execution, support, and advice for function

5. Number of persons involved in execution, support, and advice

6. Documentation forms used to support function

7. Instruments used to support function

8. Level of automation with function

9. Decision-making criteria used:
    - More problems must be dealt with
    - Diagnose or switch over to spare part

10. Priorities used within function for individual tasks

11. Source of information for function

12. Quality of information received for function

13. Destination of sending information from function
14. Quality control of outgoing information

Specific questions   None

Taking corrective actions
Generic questions

1. Scope of function
2. Time limitations of function
3. Frequency of executing function
4. Person in charge of execution, support, and advice for function
5. Number of persons involved in execution, support, and advice
6. Documentation forms used to support function
7. Instruments used to support function
8. Level of automation with function
9. Decision-making criteria used
10. Priorities used within function for individual tasks:
    - Corrective actions
    - Repair
    - Replacement
11. Source of information for function
12. Quality of information received for function
13. Destination of sending information from function
14. Quality control of outgoing information

Specific questions   None

Repairing and replacing
Generic questions

1. Scope of function
2. Time limitations of function
3. Frequency of executing function
4. Person in charge of execution, support, and advice for function
5. Number of persons involved in execution, support, and advice

6. Documentation forms used to support function

7. Instruments used to support function

8. Level of automation with function

9. Decision-making criteria used

10. Priorities used within function for individual tasks:
    - Repair criteria
    - Replacement criteria

11. Source of information for function

12. Quality of information received for function

13. Destination of sending information from function

14. Quality control of outgoing information

Specific questions

1. Tests (disruptive or nondisruptive) implemented

2. Ability to test continuously, periodically, and on-demand

Referencing to third parties

Generic questions

1. Scope of function

2. Time limitations of function

3. Frequency of executing function

4. Person in charge of execution, support, and advice for function

5. Number of persons involved in execution, support, and advice

6. Documentation forms used to support function

7. Instruments used to support function

8. Level of automation with function

9. Decision-making criteria used:
    - For referring to operations
    - For referring to vendors
    - For referring to external consultants

10. Priorities used within function for individual tasks

11. Source of information for function

12. Quality of information received for function

13. Destination of sending information from function
14. Quality control of outgoing information

Specific questions

1. List third parties
2. List any contracts or time and material

Backing up and reconfiguring
Generic questions

1. Scope of function
2. Time limitations of function
3. Frequency of executing function
4. Person in charge of execution, support, and advice for function
5. Number of persons involved in execution, support, and advice
6. Documentation forms used to support function
7. Instruments used to support function
8. Level of automation with function
9. Decision-making criteria used
10. Priorities used within function for individual tasks:
    - Decision matrix used
    - Programmed scripts used
11. Source of information for function
12. Quality of information received for function
13. Destination of sending information from function
14. Quality control of outgoing information

Specific questions

1. What components have backup
2. Who decides what components need backup

Recovering
Generic questions

1. Scope of function
2. Time limitations of function

3. Frequency of executing function

4. Person in charge of execution, support, and advice for function

5. Number of persons involved in execution, support, and advice

6. Documentation forms used to support function

7. Instruments used to support function

8. Level of automation with function

9. Decision-making criteria used

10. Priorities used within function for individual tasks

11. Source of information for function

12. Quality of information received for function

13. Destination of sending information from function

14. Quality control of outgoing information

Specific questions

1. Are tests part of recovery

Logging events and corrective actions
Generic questions

1. Scope of function

2. Time limitations of function

3. Frequency of executing function

4. Person in charge of execution, support, and advice for function

5. Number of persons involved in execution, support, and advice

6. Documentation forms used to support function

7. Instruments used to support function

8. Level of automation with function

9. Decision-making criteria used

10. Priorities used within function for individual tasks

11. Source of information for function

12. Quality of information received for function

13. Destination of sending information from function

14. Quality control of outgoing information

Specific questions

1. Any filtering used before logging

Fault Tracking
Tracking manually reported faults
Generic questions

1. Scope of function (what types of faults included)
2. Time limitations of function (lifecycle of trouble tickets)
3. Frequency of executing function:
   - Event-driven
   - Periodic for batch of problems
4. Person in charge of execution, support, and advice for function
5. Number of persons involved in execution, support, and advice
6. Documentation forms used to support function:
   - Trouble tickets with fax
   - Trouble tickets with E-mail
7. Instruments used to support function
8. Level of automation with function
9. Decision-making criteria used
10. Priorities used within function for individual tasks
11. Source of information for function
12. Quality of information received for function
13. Destination of sending information from function
14. Quality control of outgoing information

Specific questions

1. Frequency of status reports (types of faults included)
2. Automatic changes of priorities

Tracking monitored faults
Generic questions

1. Scope of function (types of faults included)
2. Time limitations of function (lifecycle of trouble tickets)

3. Frequency of executing function:
   - Event-driven
   - Periodic using SNMP-based tools
4. Person in charge of execution, support, and advice for function
5. Number of persons involved in execution, support, and advice
6. Documentation forms used to support function:
   - Trouble ticket with fax
   - Trouble ticket with E-mail
7. Instruments used to support function
8. Level of automation with function
9. Decision-making criteria used
10. Priorities used within function for individual tasks
11. Source of information for function
12. Quality of information received for function
13. Destination of sending information from function
14. Quality control of outgoing information

Specific questions

1. Polling or eventing for data collection
2. Automatic changing of priorities

Distributing information

Generic questions

1. Scope of function (information expected to be distributed)
2. Time limitations of function (trouble-ticket resolution cycles)
3. Frequency of executing function:
   - Event-driven
   - Periodic
4. Person in charge of execution, support, and advice for function
5. Number of persons involved in execution, support, and advice
6. Documentation forms used to support function
7. Instruments used to support function
8. Level of automation with function
9. Decision-making criteria used

10. Priorities used within function for individual tasks

11. Source of information for function

12. Quality of information received for function

13. Destination of sending information from function:
    - Selection by information content
    - Selection by organizational unit

14. Quality control of outgoing information

Specific questions   None

Referring

Generic questions

1. Scope of function (are all trouble tickets subject to referral)

2. Time limitations of function

3. Frequency of executing function

4. Person in charge of execution, support, and advice for function

5. Number of persons involved in execution, support, and advice

6. Documentation forms used to support function

7. Instruments used to support function

8. Level of automation with function

9. Decision-making criteria used:
   - Who is receiver
   - What trouble tickets to refer
   - After how long should trouble tickets be referred

10. Priorities used within function for individual tasks

11. Source of information for function

12. Quality of information received for function

13. Destination of sending information from function

14. Quality control of outgoing information

Specific questions

1. How to keep track of number of referrals

2. Automatic change of priorities if referral

Change Control

Managing, processing, and tracking service orders

Generic questions

1. Scope of function (definition of service orders)
2. Time limitations of function
3. Frequency of executing function:
   - Event-driven
   - Periodic
4. Person in charge of execution, support, and advice for function
5. Number of persons involved in execution, support, and advice
6. Documentation forms used to support function
7. Instruments used to support function
8. Level of automation with function
9. Decision-making criteria used:
   - Approval
   - Rejection
10. Priorities used within function for individual tasks
11. Source of information for function
12. Quality of information received for function:
    - Completeness of forms
    - Rejection criteria if not complete
13. Destination of sending information from function
14. Quality control of outgoing information

Specific questions

1. Availability of documented procedures

Routing service orders

Generic questions

1. Scope of function (clear understanding of who sends and receives service orders)
2. Time limitations of function:
   - Internal standards to support function

- Reference to organizational units and people involved in process
3. Frequency of executing function:
   - Event-driven
   - Periodic
4. Person in charge of execution, support, and advice for function
5. Number of persons involved in execution, support, and advice
6. Documentation forms used to support function
7. Instruments used to support function:
   - Instruments used
   - Applications used
8. Level of automation with function
9. Decision-making criteria used
10. Priorities used within function for individual tasks
11. Source of information for function
12. Quality of information received for function
13. Destination of sending information from function
14. Quality control of outgoing information

Specific questions

1. Process if receiver rejects service orders; any reroutes possible
2. Availability of documented procedure
3. Statistics about reroutes

Supervising the handling of changes
Generic questions

1. Scope of function (what changes included)
2. Time limitations of function:
   - Definition of various priorities
   - Commitments to priorities
3. Frequency of executing function:
   - Fixes
   - Immediate changes
   - Periodic changes

- Others

4. Person in charge of execution, support, and advice for function

5. Number of persons involved in execution, support, and advice

6. Documentation forms used to support function:
   - Supporting change process
   - Documentation of completed changes

7. Instruments used to support function

8. Level of automation with function

9. Decision-making criteria used

10. Priorities used within function for individual tasks

11. Source of information for function

12. Quality of information received for function

13. Destination of sending information from function

14. Quality control of outgoing information

Specific questions

1. Completeness of change request forms:
   - Change number
   - Date of request
   - Name of user requesting change
   - Location of user
   - Current location
   - Future location
   - Type of equipment
   - Equipment identifier
   - Target date of move
   - Actual date of move
   - Cost estimated
   - Move approved by
   - Reasons for move
   - Priority of move
   - Personnel involved
   - Fallback procedures
   - Downtime due to change
   - Notification of move to all concerned

2. Support of mass updates

3. Ratio of planned to implemented changes

Planning and Design
Analyzing needs
Generic questions

1. Scope of function (facilities and equipment considered)
2. Time limitations of function:
   - Duration of activity
   - Timeframe of need projected into future
3. Frequency of executing function:
   - On-demand
   - Periodic
4. Person in charge of execution, support, and advice for function
5. Number of persons involved in execution, support, and advice
6. Documentation forms used to support function
7. Instruments used to support function
8. Level of automation with function:
   - Manual processing of input
   - Automatic processing of input
9. Decision-making criteria used
10. Priorities used within function for individual tasks
11. Source of information for function
12. Quality of information received for function
13. Destination of sending information from function
14. Quality control of outgoing information

Specific questions

1. Differentiation by communication forms for all generic questions

Projecting application load
Generic questions

1. Scope of function (resources subject to projection)
2. Time limitations of function: 6, 12, or 18 months
3. Frequency of executing function

4. Person in charge of execution, support, and advice for function

5. Number of persons involved in execution, support, and advice

6. Documentation forms used to support function

7. Instruments used to support function

8. Level of automation with function: processing input by applications, locations, resources

9. Decision-making criteria used

10. Priorities used within function for individual tasks

11. Source of information for function

12. Quality of information received for function

13. Destination of sending information from function

14. Quality control of outgoing information

Specific questions

1. Differentiation by communications forms for all generic questions

Sizing resources

Generic questions

1. Scope of function (resources subject to sizing)

2. Time limitations of function: 6, 12, or 18 months

3. Frequency of executing function: 2 or 4 times annually

4. Person in charge of execution, support, and advice for function

5. Number of persons involved in execution, support, and advice

6. Documentation forms used to support function

7. Instruments used to support function

8. Level of automation with function

9. Decision-making criteria used

10. Priorities used within function for individual tasks

11. Source of information for function

12. Quality of information received for function

13. Destination of sending information from function

14. Quality control of outgoing information

Specific questions   None

Authorizing and tracking changes
Generic questions

1. Scope of function (resources considered for changes)
2. Time limitations of function:
   - How to handle expedited changes
   - How to handle fixes
3. Frequency of executing function:
   - On-demand
   - Periodic for large batches
4. Person in charge of execution, support, and advice for function
5. Number of persons involved in execution, support, and advice
6. Documentation forms used to support function
7. Instruments used to support function
8. Level of automation with function (evaluate consequences of changes by using configuration database)
9. Decision-making criteria used
10. Priorities used within function for individual tasks
11. Source of information for function
12. Quality of information received for function
13. Destination of sending information from function
14. Quality control of outgoing information

Specific questions

1. Is single point of contact (SPOC) used
2. How are contacts maintained with change control

Raising purchase orders
Generic questions

1. Scope of function (resources included)
2. Time limitations of function:
   - Client-driven
   - Resource-driven

3. Frequency of executing function:
   - On-demand
   - Periodic (to obtain better conditions from vendors)
4. Person in charge of execution, support, and advice for function
5. Number of persons involved in execution, support, and advice
6. Documentation forms used to support function
7. Instruments used to support function
8. Level of automation with function:
   - Evaluating orders
   - Issue orders to suppliers
9. Decision-making criteria used
10. Priorities used within this function for individual tasks (what are individual tasks)
11. Source of information for function
12. Quality of information received for function
13. Destination of sending information from function
14. Quality control of outgoing information

Specific questions

1. Is single point of contact (SPOC) used
2. Use of a checklist for order information completeness:
   - Time and date order received
   - Service order number
   - Point of contact phone and fax number
   - Work description
   - Requested completion data
   - Status of service order
   - Identification of resource
   - Building and room number
   - Ordering priority
   - Name, title, phone, and fax number of authorizing manager
   - Station number assignment
   - Actual completion date
   - Cost associated with service
3. Status update of open orders
4. Access to suppliers ordering systems via EDI or E-mail

Producing implementation plans

Generic questions

1. Scope of function (level of detail of plans)
2. Time limitations of function
3. Frequency of executing function:
   - Event-driven
   - Periodic
4. Person in charge of execution, support, and advice for function
5. Number of persons involved in execution, support, and advice
6. Documentation forms used to support function
7. Instruments used to support function
8. Level of automation with function
9. Decision-making criteria used
10. Priorities used within function for individual tasks
11. Source of information for function
12. Quality of information received for function
13. Destination of sending information from function:
    - Distribution list for primary receiver
    - Who can request copies
14. Quality control of outgoing information

Specific questions

1. What happens with returned/rejected plans

Establishing company standards

Generic questions

1. Scope of function (standards included)
2. Time limitations of function
3. Frequency of executing function:
   - On-demand
   - Periodic
4. Person in charge of execution, support, and advice for function
5. Number of persons involved in execution, support, and advice

6. Documentation forms used to support this function

7. Instruments used to support function

8. Level of automation with function

9. Decision-making criteria used

10. Priorities used within function for individual tasks

11. Source of information for function

12. Quality of information received for function

13. Destination of sending information from function

14. Quality control of outgoing information

Specific questions

1. How conformance to standards is controlled

2. How standards are considered in contracts with vendors

Maintaining quality assurance

Generic questions

1. Scope of function (what are quality standards)

2. Time limitations of function

3. Frequency of executing function:
   - On-demand
   - Periodic
   - Continuous

4. Person in charge of execution, support, and advice for function

5. Number of persons involved in execution, support, and advice

6. Documentation forms used to support function

7. Instruments used to support function

8. Level of automation with function

9. Decision-making criteria used

10. Priorities used within function for individual tasks

11. Source of information for function

12. Quality of information received for function

13. Destination of sending information from function

14. Quality control of outgoing information

Specific questions

1. Cross-functional role of this activity

Finance and Billing

Asset management

Generic questions

1. Scope of function (assets considered; how assets are defined)
2. Time limitations of function
3. Frequency of executing function:
   - On-demand updates
   - Periodic updates
4. Person in charge of execution, support, and advice for function
5. Number of persons involved in execution, support, and advice
6. Documentation forms used to support function
7. Instruments used to support function
8. Level of automation with function:
   - Processing logs
   - Processing updates
9. Decision-making criteria used:
   - Attributes included
   - Dynamic indicators included
10. Priorities used within function for individual tasks
11. Source of information for function
12. Quality of information received for function
13. Destination of sending information from function
14. Quality control of outgoing information

Specific questions

1. Existence of database (relational or object-oriented) containing records of equipment and facilities
2. Documented list of object classes and attribute classes maintained in database
3. Conformance to recommended standards
4. Level of redundancy

5. Availability of information about suppliers
6. Accessibility of addresses and names used for equipment and facilities
7. Availability of physical and logical network views
8. Status of each managed object available on demand

Costing services
Generic questions

1. Scope of function (which equipment and facilities included)
2. Time limitations of function: typical duration of activity
3. Frequency of executing function:
   - Weekly
   - Monthly
   - Quarterly
   - Semiannually
   - Annually
4. Person in charge of execution, support, and advice for function
5. Number of persons involved in execution, support, and advice
6. Documentation forms used to support function
7. Instruments used to support function
8. Level of automation with function:
   - Processing mass data
   - Distributing data
9. Decision-making criteria used and level of detail
10. Priorities used within function for individual tasks
11. Source of information for function
12. Quality of information received for function
13. Destination of sending information from function
14. Quality control of outgoing information

Specific questions

1. Fixed and operating costs of transmission and communication facilities:
   - Domestic lines
     ~Backbone networks

~Access networks
~Peer-to-peer networks

- International networks

~Backbone networks
~Access networks
~Peer-to-peer networks

- Value-added services in use

~Packet switching
~Frame relay
~Electronic mail
~LAN-to-LAN bridging service
~Managed data networks
~ISDN
~Switched digital

2. Fixed costs, depreciation, operating costs, and maintenance of communication hardware:

- Wide area networks

~Multiplexers
~Modems
~Packet switches
~Front-end processors
~Matrix switches
~Packet assembly/disassembly
~Satellite nodes (VSAT)
~Frame-relay nodes
~ATM nodes
~Concentrators
~Protocol converters
~Network management

- Local area networks

~Ethernet
~Token ring
~Token bus
~Arcnet
~Routers
~Bridges
~Extenders
~Gateways
~Network management

3. Fixed costs, depreciation, operating costs, and maintenance of communication software:

- Telecommunication access methods
- Routing software
- Operating systems
- Inventory control
- Databases
- Network management

4. Salaries, training, travel, benefits, and overhead/occupancy of personnel:

- Management
- Strategic planning
- Engineering
- Project management
- Systems software
- Operations
- Help desk
- Administration
- Product management
- External staff

5. Fixed costs, operating costs, and maintenance of infrastructure

- Space
- Cabling
- Hubs
- Power supply
- Utilities
- Security surveillance
- Backup components

6. Conformance to corporate standards

Client billing

Generic questions

1. Scope of function (components and services that can be charged back)

2. Time limitations of function

3. Frequency of executing function:

- Weekly
- Monthly
- Quarterly
- Semiannually
- Annually

4. Person in charge of execution, support, and advice for function

5. Number of persons involved in execution, support, and advice

6. Documentation forms used to support function

7. Instruments used to support function

8. Level of automation with function:
   - Data processing
   - Generating bills

9. Decision-making criteria used:
   - Level of details
   - Policy in use (proportional or responsibility)

10. Priorities used within function for individual tasks

11. Source of information for function

12. Quality of information received for function

13. Destination of sending information from function

14. Quality control of outgoing information

Specific questions

1. Conformance to corporate standards

Usage and outage collection

Generic questions

1. Scope of function (indicators in use)

2. Time limitations of function

3. Frequency of executing function

4. Person in charge of execution, support, and advice for function

5. Number of persons involved in execution, support, and advice

6. Documentation forms used to support function

7. Instruments used to support function

8. Level of automation with function

9. Decision-making criteria used

10. Priorities used within function for individual tasks

11. Source of information for function

12. Quality of information received for function

13. Destination of sending information from function

14. Quality control of outgoing information

Specific questions   None

Calculation of rebates to clients
Generic questions

1. Scope of function (magnitude of rebates)

2. Time limitations of function

3. Frequency of executing function

4. Person in charge of execution, support, and advice for function

5. Number of persons involved in execution, support, and advice

6. Documentation forms used to support function

7. Instruments used to support function

8. Level of automation with function:

   - Criteria when rebates are due
   - Preprogrammed criteria; scanning of files and service-level agreements

9. Decision-making criteria used

10. Priorities used within function for individual tasks

11. Source of information for function

12. Quality of information received for function

13. Destination of sending information from function

14. Quality control of outgoing information

Specific questions

1. Is reimbursement desired in cases of noncompliance

2. Any special deals with clients

Bill verification
Generic questions

1. Scope of function

2. Time limitations of function:

   - In concert with payments due
   - In concert with charging due

3. Frequency of executing function:
   - On-demand
   - Periodic
4. Person in charge of execution, support, and advice for function
5. Number of persons involved in execution, support, and advice
6. Documentation forms used to support function
7. Instruments used to support function
8. Level of automation with function:
   - Comparison of bills
   - Scanning for unusual exceptions
9. Decision-making criteria used:
   - Sampling by vendors, facilities, equipment
   - In-depth for all vendors, facilities, and equipment
10. Priorities used within function for individual tasks
11. Source of information for function
12. Quality of information received for function
13. Destination of sending information from function
14. Quality control of outgoing information

Specific questions

1. Communication forms separate or together

Software license control

Generic questions

1. Scope of function
2. Time limitations of function
3. Frequency of executing function:
   - Sampling
   - Periodic
4. Person in charge of execution, support, and advice for function
5. Number of persons involved in execution, support, and advice
6. Documentation forms used to support function
7. Instruments used to support function

8. Level of automation with function and evaluation of counters

9. Decision-making criteria used:
   - Information used for accounting
   - Information used for usage confirmation
   - Both

10. Priorities used within function for individual tasks

11. Source of information for function

12. Quality of information received for function

13. Destination of sending information from function

14. Quality control of outgoing information

Specific questions

1. Products and vendors considered

Implementation and Maintenance

Implementing change requests and work orders

Generic questions

1. Scope of function (changes and work orders considered)

2. Time limitations of function

3. Frequency of executing function

4. Person in charge of execution, support, and advice for function

5. Number of persons involved in execution, support, and advice

6. Documentation forms used to support function

7. Instruments used to support function

8. Level of automation with function

9. Decision-making criteria used

10. Priorities used within function for individual tasks and which tasks supported

11. Source of information for function

12. Quality of information received for function

13. Destination of sending information from function

14. Quality control of outgoing information

Specific questions

1. Availability of procedures for mass updates after successful provisioning and changes
2. Availability of testing procedures

Maintaining resources

Generic questions

1. Scope of function:
   - Resources considered for maintenance
   - What maintenance includes
   - Any backups considered for maintenance
2. Time limitations of function; typical duration of maintenance activities
3. Frequency of executing function
   - On-demand
   - Preventive
4. Person in charge of execution, support, and advice for function
5. Number of persons involved in execution, support, and advice
6. Documentation forms used to support function
7. Instruments used to support function
8. Level of automation with function:
   - Evaluation of logs
   - Evaluation of maintenance records of facilities and equipment
9. Decision-making criteria used for preventive and on-demand maintenance
10. Priorities used within function for individual tasks
11. Source of information for function
12. Quality of information received for function
13. Destination of sending information from function
14. Quality control of outgoing information

Specific questions

1. Any considerations for outsourcing maintenance

Inspection

Generic questions

1. Scope of function
   - Resources considered for maintenance
   - What maintenance includes
   - Any backups considered for maintenance
2. Time limitations of function; typical duration of inspection activities
3. Frequency of executing function:
   - Preventive
   - On-demand
4. Person in charge of execution, support, and advice for function
5. Number of persons involved in execution, support, and advice
6. Documentation forms used to support function
7. Instruments used to support function
8. Level of automation with function:
   - Evaluation of logs
   - Evaluation of inspection results
9. Decision-making criteria used
   - On-demand
   - Preventive inspection
10. Priorities used within function for individual tasks
11. Source of information for function
12. Quality of information received for function
13. Destination of sending information from function
14. Quality control of outgoing information

Specific questions

1. Consideration for outsourcing
2. Does environmental control meet requirements in terms of heating ventilation air conditioning (HVAC) systems and power backups
3. Availability of specific power backup using uninterrupted power supply (UPS)
4. Priorities defined for power backups

5. Alternative sites defined for major outages

6. Comfortable, ergonomic furnishings

7. Availability of intersite communication media:
   - Phones, fax, telex
   - E-mail
   - Closed-circuit television
   - Video

8. Protection of sensitive areas

Maintaining configuration database

Generic questions

1. Scope of function (managed objects included)

2. Time limitations of function

3. Frequency of executing function:
   - Immediate updates
   - Delayed mass updates

4. Person in charge of execution, support, and advice for function

5. Number of persons involved in execution, support, and advice

6. Documentation forms used to support function

7. Instruments used to support function

8. Level of automation with function; evaluation of fixes and changes that need updates

9. Decision-making criteria used; single or batch updates

10. Priorities used within function for individual tasks

11. Source of information for function

12. Quality of information received for function

13. Destination of sending information from function

14. Quality control of outgoing information

Specific questions

1. Availability of current network configuration

2. Availability of past network configurations

3. Availability of layered topology displays

4. Online record of currently installed equipment and facilities, including spares

5. Availability of detailed information about connectivity

6. Status of each managed object available on-demand

7. Generation of physical and logical views supported

8. Sectionalization of the network supported

Provisioning

Generic questions

1. Scope of function (managed objects subject to provisioning)

2. Time limitations of function

3. Frequency of executing function:
   - On-demand
   - Periodic with larger batches from same supplier

4. Person in charge of execution, support, and advice for this function

5. Number of persons involved in execution, support, and advice

6. Documentation forms used to support function

7. Instruments used to support function

8. Level of automation with function

9. Decision-making criteria used

10. Priorities used within function for individual tasks

11. Source of information for function:
    - Use of EDI
    - Use of other communication forms

12. Quality of information received for function

13. Destination of sending information from function

14. Quality control of outgoing information

Specific questions

1. Deadline guarantees from suppliers

2. Penalties for noncompliance

Fault Monitoring

Monitoring system and network for proactive problem detection

Generic questions

1. Scope of function (facilities and equipment monitored)
2. Time limitations of function
3. Frequency of executing function:
   - Event-driven (CMIP-type protocols)
   - Periodic (SNMP-type protocols)
4. Person in charge of execution, support, and advice for function
5. Number of persons involved in execution, support, and advice
6. Documentation forms used to support function
7. Instruments used to support function:
   - Data collection
   - Interpretation
   - Visualization
8. Level of automation with function:
   - Data collection
   - Interpretation
   - Visualization
9. Decision-making criteria used:
   - Level of details for displays
   - Level of details for logs
10. Priorities used within function for individual tasks
11. Source of information for function
12. Quality of information received for function
13. Destination of sending information from function
14. Quality control of outgoing information

Specific questions

1. Supervision of systems and networks status by indicators
2. Capabilities of correlating alarms:
   - Automated alarm correlation
   - Use of expert systems
3. Capabilities of event notification:

- Events
- Alarm generation
- Alarm correlation
- Message filtering

4. Managed objects that offer built-in monitoring capabilities:

  - Communication software

  - WAN elements (modems, multiplexers, packet switches, matrix switches, frame-relay switches, ATM switches, cellular switches, etc.)

  - LAN elements (LAN segments, routers, gateways, bridges, extenders, etc.)

  - MAN elements (FDDI nodes, DQDB nodes, etc.)

  - PBXs

  - Others

5. Problem detection:

  - Network status changes updated on real-time basis
  - Real-time statistics available
  - Contact index with names and phone numbers available

Opening additional trouble tickets

Generic questions

1. Scope of function

2. Time limitations of function

3. Frequency of executing function

4. Person in charge of execution, support, and advice for function

5. Number of persons involved in execution, support, and advice

6. Documentation forms used to support function

7. Instruments used to support function

8. Level of automation with function

9. Decision-making criteria used

10. Priorities used within function for individual tasks

11. Source of information for function:

  - Information provided manually
  - Information provided automatically

12. Quality of information received for function

13. Destination of sending information from function
14. Quality control of outgoing information

Specific questions

1. Checking open trouble tickets
2. Checking closed trouble tickets
3. Checking history files of trouble tickets

Referring trouble tickets
Generic questions

1. Scope of function (referring criteria)
2. Time limitations of function
3. Frequency of executing function
4. Person in charge of execution, support, and advice for function
5. Number of persons involved in execution, support, and advice
6. Documentation forms used to support function
7. Instruments used to support function
8. Level of automation with function
9. Decision-making criteria used:
   - Who sets referring criteria
   - What are escalation timeframes
10. Priorities used within function for individual tasks:
    - By managed objects
    - By duration
    - By severity of problems
11. Source of information for function
12. Quality of information received for function
13. Destination of sending information from function
14. Quality control of outgoing information

Specific questions

1. Availability of statistics about referrals
2. What happens if referral is rejected

Performance Monitoring

Monitoring system and networks performance

Generic questions

1. Scope of function (indicators considered):
   - Workload indicators
   - Service indicators
   - Throughput indicators
   - Utilization indicators

2. Time limitations of function; deadlines if performance bottle-necks discovered

3. Frequency of executing function:
   - Continuous
   - Periodic

4. Person in charge of execution, support, and advice for function

5. Number of persons involved in execution, support, and advice

6. Documentation forms used to support function

7. Instruments used to support function

8. Level of automation with function

9. Decision-making criteria used:
   - Indicators considered
   - When to change resolution levels
   - What to display

10. Priorities used within function for individual tasks

11. Source of information for function

12. Quality of information received for function

13. Destination of sending information from function

14. Quality control of outgoing information

Specific questions   None

Monitoring service-level agreements

Generic questions

1. Scope of function (indicators included for SLAs):
   - Contracting parties
   - Expiration date of agreement

- Procedures for modification including authorization
- Workload description and quantification
- Service-level indicators and quantification
- Determination of responsibilities
- Reporting procedures and intervals
- Accounting procedures and intervals
- Monetary charges and reimbursement for noncompliance
- Escalation procedures if deviation from agreement
- Contact names, addresses, phone, fax numbers of responsible persons

2. Time limitations of function
3. Frequency of executing function:

- Continuous
- Periodic
- On-demand

4. Person in charge of execution, support, and advice for function
5. Number of persons involved in execution, support, and advice
6. Documentation forms used to support function
7. Instruments used to support function
8. Level of automation with function:

- Automatic compression
- Automatic consolidation

9. Decision-making criteria used and escalation for noncompliance
10. Priorities used within function for individual tasks
11. Source of information for function
12. Quality of information received for function
13. Destination of sending information from function
14. Quality control of outgoing information

Specific questions

1. What reports are generated

Monitoring third-party and vendor performance
Generic questions

1. Scope of function (what indicators used)
2. Time limitations of function

3. Frequency of executing function:

- Continuous
- Periodic
- On-demand

4. Person in charge of execution, support, and advice for function

5. Number of persons involved in execution, support, and advice

6. Documentation forms used to support function

7. Instruments used to support function

8. Level of automation with function:

- Automated compression
- Automated consolidation

9. Decision-making criteria used

10. Priorities used within function for individual tasks

11. Source of information for function

12. Quality of information received for function

13. Destination of sending information from function

14. Quality control of outgoing information

Specific questions

1. What reports are generated

Optimization, modeling, and tuning
Generic questions

1. Scope of function (indicators and managed objects considered)

2. Time limitations of function:

- Permanent function
- Mission-driven function

3. Frequency of executing function:

- Permanent function
- Mission-driven function

4. Person in charge of execution, support, and advice for function

5. Number of persons involved in execution, support, and advice

6. Documentation forms used to support function

7. Instruments used to support function

8. Level of automation with function:

- Extraction from database
- Processing measurement data

9. Decision-making criteria used and feasibility of hypothesis
10. Priorities used within function for individual tasks
11. Source of information for function
12. Quality of information received for function
13. Destination of sending information from function
14. Quality control of outgoing information

Specific questions

1. Is baselining supported
2. Any preference toward analytic or simulative techniques

Reporting usage statistics and trends
Generic questions

1. Scope of function (performance data reported)
2. Time limitations of function; time gap for processing data
3. Frequency of executing function:
   - On-demand
   - Periodic: daily, weekly, monthly, quarterly, etc.
4. Person in charge of execution, support, and advice for function
5. Number of persons involved in execution, support, and advice
6. Documentation forms used to support function
7. Instruments used to support function
8. Level of automation with function:
   - Preprocessing of data
   - Report generation
9. Decision-making criteria used (thresholding)
10. Priorities used within function for individual tasks
11. Source of information for function
12. Quality of information received for function
13. Destination of sending information from function
14. Quality control of outgoing information:

- Type of reports, such as pie charts, tables, graphics, others
- Thresholds crossed
- Trends in outages, utilization
- Error levels and frequency distribution
- Analysis of problems

Specific questions

1. What are preferred presentation forms

Security Management

Threat analysis

Generic questions

1. Scope of function (managed objects included)
2. Time limitations of function:
   - How long analysis takes under stress
   - How long analysis takes under no stress
3. Frequency of executing function
4. Person in charge of execution, support, and advice for function
5. Number of persons involved in execution, support, and advice
6. Documentation forms used to support function
7. Instruments used to support function
8. Level of automation with function (does not apply)
9. Decision-making criteria used:
   - Setting priorities
   - Defining actions against intruder
10. Priorities used within function for individual tasks
11. Source of information for function
12. Quality of information received for function
13. Destination of sending information from function
14. Quality control of outgoing information

Specific questions

1. Availability of violation statistics
2. Security management team

Administration

Generic questions

1. Scope of function (managed objects included in security administration)
2. Time limitations of function
3. Frequency of executing function
4. Person in charge of execution, support, and advice for function
5. Number of persons involved in execution, support, and advice
6. Documentation forms used to support function
7. Instruments used to support function
8. Level of automation with function
9. Decision-making criteria used
10. Priorities used within function for individual tasks
11. Source of information for function
12. Quality of information received for function
13. Destination of sending information from function
14. Quality control of outgoing information

Specific questions   None

Detection

Generic questions

1. Scope of function (security indicators considered for managed objects)
2. Time limitations of function; time intervals available for detection
3. Frequency of executing function
4. Person in charge of execution, support, and advice for function
5. Number of persons involved in execution, support, and advice
6. Documentation forms used to support function
7. Instruments used to support function
8. Level of automation with function
9. Decision-making criteria used:
   - Thresholds used for severity
   - Actions taken by severity

10. Priorities used within function for individual tasks

11. Source of information for function

12. Quality of information received for function

13. Destination of sending information from function

14. Quality control of outgoing information

Specific questions   None

Recovery

Generic questions

1. Scope of function (managed objects considered)

2. Time limitations of function; usual recovery deadlines

3. Frequency of executing function:
   - Periodic
   - On-demand

4. Person in charge of execution, support, and advice for function

5. Number of persons involved in execution, support, and advice

6. Documentation forms used to support function

7. Instruments used to support function

8. Level of automation with function

9. Decision-making criteria used

10. Priorities used within function for individual tasks

11. Source of information for function

12. Quality of information received for function

13. Destination of sending information from function

14. Quality control of outgoing information

Specific questions

1. Availability of recovery procedures

Protecting the management systems

Generic questions

1. Scope of function (solutions for authorization and authentication)

2. Time limitations of function:

- Detection of problems
- Actions taken

3. Frequency of executing function:
   - Sampling
   - Periodic

4. Person in charge of execution, support, and advice for function

5. Number of persons involved in execution, support, and advice

6. Documentation forms used to support function

7. Instruments used to support function

8. Level of automation with function:
   - Detection of violations
   - Actions taken against intruder or violator

9. Decision-making criteria used

10. Priorities used within function for individual tasks

11. Source of information for function

12. Quality of information received for function

13. Destination of sending information from function

14. Quality control of outgoing information

Specific questions

1. Availability of more detailed description of security features of network management system

2. Partitioning capabilities

3. Security management solutions in place:
   - Encryption
   - Digital signature
   - Chip keys or chip cards
   - Physical access control
   - Passwords

Systems Administration

Software version control
Generic questions

1. Scope of function

2. Time limitations of function

3. Frequency of executing function:
   - Sampling
   - Periodic
4. Person in charge of execution, support, and advice for function
5. Number of persons involved in execution, support, and advice
6. Documentation forms used to support function
7. Instruments used to support function
8. Level of automation with function; automatic extraction of inventory data
9. Decision-making criteria used; what and when to upgrade
10. Priorities used within function for individual tasks
11. Source of information for function
12. Quality of information received for function
13. Destination of sending information from function
14. Quality control of outgoing information

Specific questions

1. How information is received from remote sites

Software distribution
Generic questions

1. Scope of function
2. Time limitations of function
3. Frequency of executing function:
   - Push
   - Pull
4. Person in charge of execution, support, and advice for function
5. Number of persons involved in execution, support, and advice
6. Documentation forms used to support function
7. Instruments used to support function
8. Level of automation with function:
   - Status supervision
   - Controlling the distribution process
9. Decision-making criteria used; backup when distribution fails

10. Priorities used within function for individual tasks
11. Source of information for function
12. Quality of information received for function
13. Destination of sending information from function
14. Quality control of outgoing information

Specific questions

1. Combination with software licensing

Systems management
Generic questions

1. Scope of function (systems and indicators considered)
2. Time limitations of function; supervision cycles to evaluate status
3. Frequency of executing function:
   - Polling
   - Eventing
4. Person in charge of execution, support, and advice for function
5. Number of persons involved in execution, support, and advice
6. Documentation forms used to support function
7. Instruments used to support function
8. Level of automation with function:
   - Supervision status
   - Processing data
   - Reporting
9. Decision-making criteria used:
   - Choice of indicators
   - Choice of reporting intervals
   - Centralization or decentralization
   - Inband or outband concept
10. Priorities used within function for individual tasks
11. Source of information for function
12. Quality of information received for function
13. Destination of sending information from function
14. Quality control of outgoing information

Specific questions   None

Administration of user-definable tables
Generic questions

1. Scope of function (data included)
2. Time limitations of function; duration of typical updates
3. Frequency of executing function
4. Person in charge of execution, support, and advice for function
5. Number of persons involved in execution, support, and advice
6. Documentation forms used to support function
7. Instruments used to support function
8. Level of automation with function:
   - Automated updates
   - Mass updates
9. Decision-making criteria used; person authorized to make changes
10. Priorities used within function for individual tasks
11. Source of information for function
12. Quality of information received for function
13. Destination of sending information from function
14. Quality control of outgoing information

Specific questions

1. Pointers to other databases

Local and remote configuration of resources
Generic questions

1. Scope of function (resources considered)
2. Time limitations of function:
   - How much time available when interruption required
   - Time limits for uninterrupted configuring
3. Frequency of executing function:
   - On-demand
   - Periodic

4. Person in charge of execution, support, and advice for function

5. Number of persons involved in execution, support, and advice

6. Documentation forms used to support function

7. Instruments used to support function

8. Level of automation with function:
   - Measuring status
   - Interpreting status

9. Decision-making criteria used:
   - Indicators to configure
   - When to configure

10. Priorities used within function for individual tasks:
    - Local first
    - Remote locations first

11. Source of information for function

12. Quality of information received for function

13. Destination of sending information from function

14. Quality control of outgoing information

Specific questions

1. Pointers to configuration database(s)

Management of names and addresses

Generic questions

1. Scope of function (distribution of addresses and names)

2. Time limitations of function; duration of typical changes with names and with addresses

3. Frequency of executing function:
   - On-demand
   - Periodic

4. Person in charge of execution, support, and advice for function

5. Number of persons involved in execution, support, and advice

6. Documentation forms used to support function

7. Instruments used to support function

8. Level of automation with function:

- Checking duplications
- Checking alias names

9. Decision-making criteria used; lengths and structures of names and addresses

10. Priorities used within function for individual tasks

11. Source of information for function

12. Quality of information received for function

13. Destination of sending information from function

14. Quality control of outgoing information

Specific questions

1. Use of Internet for commercial purposes

Applications management

Generic questions

1. Scope of function (applications and databases included)

2. Time limitations of function

3. Frequency of executing function:
   - Polling
   - Eventing

4. Person in charge of execution, support, and advice for function

5. Number of persons involved in execution, support, and advice

6. Documentation forms used to support function

7. Instruments used to support function

8. Level of automation with function:
   - Supervision of status
   - Processing data
   - Reporting

9. Decision-making criteria used:
   - Choice of indicators
   - Choice of reports
   - Centralized or decentralized management
   - Outband or inband concept

10. Priorities used within function for individual tasks

11. Source of information for function

12. Quality of information received for function
13. Destination of sending information from function
14. Quality control of outgoing information

Specific questions   None

Bibliography

Compass America, Inc., *Network Compass Users Guide, Compass America Incorporated*, Herndon, VA, 1993.

Herman, J., *Strategic Planning for Distributed Systems Management, Tutorial*, Network and Distributed Systems Management, Wash. D.C., *Technology Transfer Institute,* Conference Record, 1994.

Hutchins, G., *The ISO-9000 Implementation Manual*, Omneo, *Oliver Wright Publication*, Essex Junction, NJ, 1994.

Network Management Forum, *Discovering OMNIPoint, PTR Prentice Hall*, Englewood Cliffs, NJ, 1993.

Network Management Forum, *Statement of User Requirements for Management of Networked Information Systems*, Network Management Forum, Morristown, NJ, 1992.

Real Decisions Corp., *Comprehensive Voice and Data Network Analysis, Real Decisions Corporation*, Darien, CT, 1993.

Terplan, Kornel, *Communication Networks Management, Prentice Hall*, Second Edition, Englewood Cliffs, NJ, 1992.

_____ *Effective Management of Local Area Networks—Functions, Instruments, and People, McGraw-Hill, Inc.*, NY, 1992.

_____ Huntington-Lee, J., *Applications for Distributed Systems and Network Management, Van Nostrand and Reynolds*, NY, 1994.

The Yankee Group, *Global Network Strategies Conference, The Yankee Group*, New York, October 13–14, 1994.

Shrewsbury, J. K., *Telecommunications Management Network (TMN)*, Technology Survey, *WilTel, Inc.*, Tulsa, OK, 1994.

Index

Illustrations are in **boldface**

ABOUT THE AUTHOR

Kornel Terplan is a telecommunications expert with more than 25 years of corporate consulting experience. He has provided consulting, training, and product development services to more than 75 national and multinational corporations on four continents, including AT&T, GTE, Walt Disney World, BMW, and Siemens. Dr. Terplan is the author of more than 140 articles and 10 books, including *Communication Networks Management* and *Effective Management of Local Area Networks*. He received his doctoral degree from the University of Dresden, Germany.